Teacher Edition

Eureka Math® Grade K Module 3

Special thanks go to the Gordon A. Cain Center and to the Department of Mathematics at Louisiana State University for their support in the development of *Eureka Math*.

For a free *Eureka Math* Teacher Resource Pack, Parent Tip Sheets, and more please visit
https://eurekamath.greatminds.org/teacher-resource-pack

Published by the Great Minds

Copyright © 2015 Great Minds®. No part of this work may be reproduced, sold, or commercialized, in whole or in part, without written permission from Great Minds. Non-commercial use is licensed pursuant to a Creative Commons Attribution-NonCommercial-ShareAlike 4.0 license; for more information, go to http://greatminds.net/maps/math/copyright. "Great Minds" and "Eureka Math" are registered trademarks of Great Minds.

Printed in the U.S.A.

This book may be purchased from the publisher at eureka-math.org

BAB 10 9 8 7 6 5 4 3

ISBN 978-1-63255-343-0

Eureka Math: A Story of Units® **Contributors**

Katrina Abdussalaam, Curriculum Writer
Tiah Alphonso, Program Manager—Curriculum Production
Kelly Alsup, Lead Writer / Editor, Grade 4
Catriona Anderson, Program Manager—Implementation Support
Debbie Andorka-Aceves, Curriculum Writer
Eric Angel, Curriculum Writer
Leslie Arceneaux, Lead Writer / Editor, Grade 5
Kate McGill Austin, Lead Writer / Editor, Grades PreK–K
Adam Baker, Lead Writer / Editor, Grade 5
Scott Baldridge, Lead Mathematician and Lead Curriculum Writer
Beth Barnes, Curriculum Writer
Bonnie Bergstresser, Math Auditor
Bill Davidson, Fluency Specialist
Jill Diniz, Program Director
Nancy Diorio, Curriculum Writer
Nancy Doorey, Assessment Advisor
Lacy Endo-Peery, Lead Writer / Editor, Grades PreK–K
Ana Estela, Curriculum Writer
Lessa Faltermann, Math Auditor
Janice Fan, Curriculum Writer
Ellen Fort, Math Auditor
Peggy Golden, Curriculum Writer
Maria Gomes, Pre-Kindergarten Practitioner
Pam Goodner, Curriculum Writer
Greg Gorman, Curriculum Writer
Melanie Gutierrez, Curriculum Writer
Bob Hollister, Math Auditor
Kelley Isinger, Curriculum Writer
Nuhad Jamal, Curriculum Writer
Mary Jones, Lead Writer / Editor, Grade 4
Halle Kananak, Curriculum Writer
Susan Lee, Lead Writer / Editor, Grade 3
Jennifer Loftin, Program Manager—Professional Development
Soo Jin Lu, Curriculum Writer
Nell McAnelly, Project Director

Ben McCarty, Lead Mathematician / Editor, PreK–5
Stacie McClintock, Document Production Manager
Cristina Metcalf, Lead Writer / Editor, Grade 3
Susan Midlarsky, Curriculum Writer
Pat Mohr, Curriculum Writer
Sarah Oyler, Document Coordinator
Victoria Peacock, Curriculum Writer
Jenny Petrosino, Curriculum Writer
Terrie Poehl, Math Auditor
Robin Ramos, Lead Curriculum Writer / Editor, PreK–5
Kristen Riedel, Math Audit Team Lead
Cecilia Rudzitis, Curriculum Writer
Tricia Salerno, Curriculum Writer
Chris Sarlo, Curriculum Writer
Ann Rose Sentoro, Curriculum Writer
Colleen Sheeron, Lead Writer / Editor, Grade 2
Gail Smith, Curriculum Writer
Shelley Snow, Curriculum Writer
Robyn Sorenson, Math Auditor
Kelly Spinks, Curriculum Writer
Marianne Strayton, Lead Writer / Editor, Grade 1
Theresa Streeter, Math Auditor
Lily Talcott, Curriculum Writer
Kevin Tougher, Curriculum Writer
Saffron VanGalder, Lead Writer / Editor, Grade 3
Lisa Watts-Lawton, Lead Writer / Editor, Grade 2
Erin Wheeler, Curriculum Writer
MaryJo Wieland, Curriculum Writer
Allison Witcraft, Math Auditor
Jessa Woods, Curriculum Writer
Hae Jung Yang, Lead Writer / Editor, Grade 1

Board of Trustees

Lynne Munson, President and Executive Director of Great Minds
Nell McAnelly, Chairman, Co-Director Emeritus of the Gordon A. Cain Center for STEM Literacy at Louisiana State University
William Kelly, Treasurer, Co-Founder and CEO at ReelDx
Jason Griffiths, Secretary, Director of Programs at the National Academy of Advanced Teacher Education
Pascal Forgione, Former Executive Director of the Center on K-12 Assessment and Performance Management at ETS
Lorraine Griffith, Title I Reading Specialist at West Buncombe Elementary School in Asheville, North Carolina
Bill Honig, President of the Consortium on Reading Excellence (CORE)
Richard Kessler, Executive Dean of Mannes College the New School for Music
Chi Kim, Former Superintendent, Ross School District
Karen LeFever, Executive Vice President and Chief Development Officer at ChanceLight Behavioral Health and Education
Maria Neira, Former Vice President, New York State United Teachers

A STORY OF UNITS

Mathematics Curriculum

GRADE K • MODULE 3

Table of Contents
GRADE K • MODULE 3
Comparison of Length, Weight, Capacity, and Numbers to 10

Module Overview 2

Topic A: Comparison of Length and Height 10

Topic B: Comparison of Length and Height of Linking Cube Sticks Within 10 38

Topic C: Comparison of Weight 77

Topic D: Comparison of Volume 112

Mid-Module Assessment and Rubric 149

Topic E: Are There Enough? 158

Topic F: Comparison of Sets Within 10 207

Topic G: Comparison of Numerals 246

Topic H: Clarification of Measurable Attributes 279

End-of-Module Assessment and Rubric 307

Answer Key 315

A STORY OF UNITS   Module Overview   K•3

Grade K • Module 3

Comparison of Length, Weight, Capacity, and Numbers to 10

OVERVIEW

Having observed, analyzed, and classified objects by shape into predetermined categories in Module 2, students now compare and analyze length, weight, capacity, and finally, numbers in Module 3. Students use language such as *longer than, shorter than, as long as; heavier than, lighter than, as heavy as;* and *more than, less than, the same as.* "8 is *more than* 5." "5 is *less than* 8." "5 is *the same as* 5." "2 and 3 is also *the same as* 5."

Topics A and B focus on comparison of length, Topic C on comparison of weight, and Topic D on comparison of volume (**K.MD.2**). Each of these topics opens with an identification of the attribute being compared within the natural context of the lesson (**K.MD.1**). For example, in Topic A, before exploring length, students realize they could have chosen to compare by a different attribute: weight, length, volume, or numbers (**K.MD.1**).

    T: Students, when you compare and say it is bigger, let's think about what you mean. (After each question, allow students to have a lively, brief discussion.)

    T: Do you mean that it is bigger, like this book is *heavier than* this ribbon? (Dramatize the weight of the book and ribbon.)

    T: Do you mean that it is longer, like this ribbon is *longer than* this book? (Dramatize the length of the ribbon.)

    T: Do you mean that it takes up more space, like this book *takes up more space* than this ribbon when it is all squished together? (Dramatize.)

    T: Do you mean to compare the number of things, like *the number* of books and ribbons? (Dramatize a count.)

    T: So, we can compare things in different ways! Today, let's compare by thinking about longer than, taller than, or shorter than. (Dramatize.)

After the Mid-Module Assessment, Topic E begins with an analysis using the question, "Are there enough?" This leads naturally from exploring when and if there is enough space to seeing whether there are enough chairs for a small set of students: "There are fewer chairs than students!" This bridges into Topics F and G, which present a sequence building toward the comparison of numerals (**K.CC.7**). Topic F begins with counting and matching sets to compare (**K.CC.6**). The module culminates in a three-day exploration, one day devoted to each attribute: length, weight, and volume (**K.MD.2**). The module closes with a culminating task devoted to distinguishing between the measurable attributes of a set of objects: a water bottle, cup, dropper, and juice box (**K.MD.1**).

2   Module 3:   Comparison of Length, Weight, Capacity, and Numbers to 10

© 2015 Great Minds. eureka-math.org
GK-M3-TE-B3-1.3.1-01.2016

A STORY OF UNITS — Module Overview — K•3

The module supports students' understanding of amounts and their developing number sense. For example, counting how many small cups of rice are contained within a larger quantity provides a foundational concept of place value: Within a larger amount are smaller equal units, which together make up the whole. "4 cups of rice is the same as 1 mug of rice." Compare that statement to "10 ones is the same as 1 ten" (**1.NBT.2a**). As students become confident directly comparing the length of a pencil and a crayon with statements such as "The pencil is longer than the crayon" (**K.MD.2**), they will be ready in later grades to indirectly compare using length units with statements such as "The pencil is longer than the crayon because 7 cubes is more than 4 cubes" (**1.MD.2**).

Additional foundational work for later grades is as follows:

- **Foundational work with equivalence.** The length of a stick with 5 linking cubes is the same as the length of my cell phone. A pencil weighs the same as a stick with 5 linking cubes. Each module component on measurement closes with a focus on *the same as*.

- **Foundational work for the precise use and understanding of rulers and number lines.** The module opens with lessons pointing out the importance of aligning endpoints to measure length.

- **Foundational understanding of area.** At the opening of the second half of the module, students informally explore area as they see whether a yellow circle fits inside a red square. They then see how many small blue squares will fit inside the red square and, finally, that many beans will cover the same area (pictured to the right).

- **Foundational understanding of comparison.** As students count to compare the length of linking cube sticks, they are laying the foundation for answering *how many more…than/less…than* questions in Grade 1 (**1.MD.2**).

Notes on Pacing for Differentiation

Sprints are introduced in the second half of this module through a gradual progression of preparation exercises. When consolidating or omitting lessons, take care to maintain the intended sequence of the Sprints as listed.

Consider omitting Lesson 7. In order to do so, offer *the same as* as one more option to describe the comparison in Lessons 4–6. Be sure to include objects for comparison that yield descriptions of *shorter than*, *longer than*, and *the same length as*.

If students progress quickly in comparing weight by estimating, they may be ready to use the balance scale sooner, allowing for the consolidation of Lessons 8 and 9. To bridge their understanding, have students model the movement of the balance scale with their arms and hands.

Students might better grasp the concepts of volume and capacity if they observe first and explore afterwards. Consider consolidating Lessons 13–15 into a series of demonstrations with students engaged chorally, as recorders, and as acute observers (e.g., "Count the scoops as I fill the container"; "Record the number of scoops it took to fill the container"; and "Share with your partner about what happened to the water"). Students might then gain hands-on experience and explore the concept later (e.g., in centers, science). If pacing is a challenge and students study volume as part of science, consider omitting Lessons 14 and 15.

Module 3: Comparison of Length, Weight, Capacity, and Numbers to 10

Consider omitting Lesson 16; although engaging and interesting, students may not need the introduction to area through informal comparison.

Topic H serves as a culminating topic where students synthesize their knowledge of the attributes previously studied in this module. Because no new learning is introduced, these lessons might be omitted or moved to another time of day.

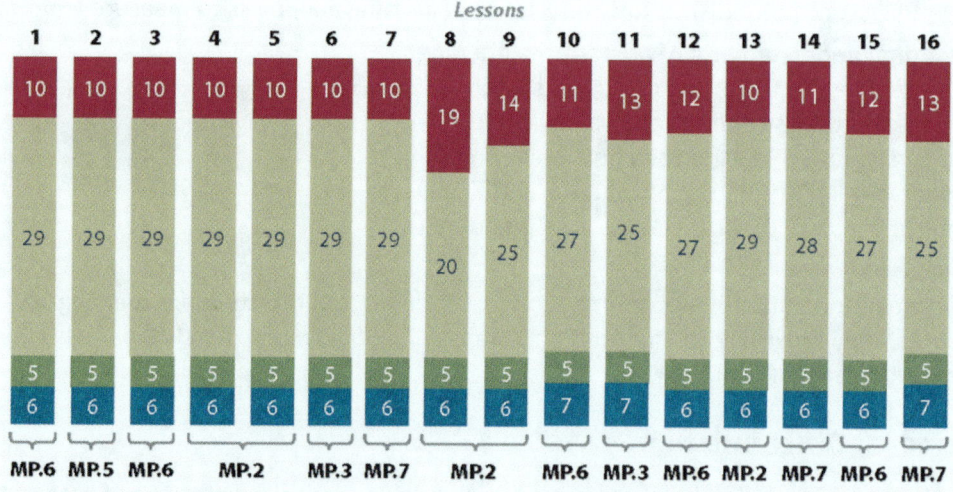

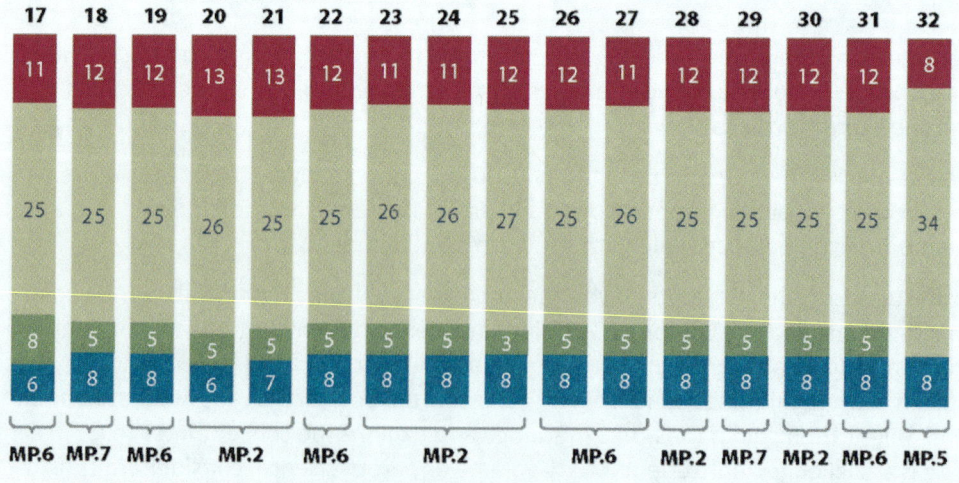

Focus Grade Level Standards

Compare numbers.

K.CC.6   Identify whether the number of objects in one group is greater than, less than, or equal to the number of objects in another group, e.g., by using matching and counting strategies. (Include groups with up to 10 objects.)

K.CC.7   Compare two numbers between 1 and 10 presented as written numerals.

Describe and compare measurable attributes.

K.MD.1   Describe measurable attributes of objects, such as length or weight. Describe several measurable attributes of a single object.

K.MD.2   Directly compare two objects with a measurable attribute in common, to see which object has "more of"/"less of" the attribute, and describe the difference. *For example, directly compare the heights of two children and describe one child as taller/shorter.*

Foundational Standards

PK.CC.5   Identify whether the number of objects in one group is more, less, greater than, fewer, and/or equal to the number of objects in another group, e.g., by using matching and counting strategies.[1]

PK.CC.6   Identify "first" and "last" related to order or position.

PK.MD.1   Identify measurable attributes of objects, such as length and weight. Describe them using correct vocabulary (e.g., small, big, short, tall, empty, full, heavy, and light).

Focus Standards for Mathematical Practice

MP.2   **Reason quantitatively and abstractly.** Students compare quantities by drawing objects in columns and matching the objects one to one to see that one column has more than another and draw the conclusion that 6 is more than 4 because 2 objects do not have a match.

MP.3   **Construct viable arguments and critique the reasoning of others.** Students describe measurable attributes of a single object and reason about how to compare its length, weight, and volume to that of another object.

MP.5   **Use appropriate tools strategically.** During the culminating task and End-of-Module Assessment, students might choose to use a scale to compare weight, linking cube sticks to compare length and rice and cups to compare volume.

---

[1] Up to 5 objects

A STORY OF UNITS — Module Overview — K•3

MP.6 **Attend to precision.** Students attend to precision by aligning endpoints when comparing lengths. They are also precise when weighing an object with cubes (or units) on a balance scale. Adding 1 more makes the cubes too heavy when the goal is to see how many cubes have the same weight as the object.

MP.7 **Look for and make use of structure.** Students use structure to see that the amount of rice in 1 container is equal to the amount in 4 smaller containers. The smaller unit is a structure, as is the larger unit.

Overview of Module Topics and Lesson Objectives

Standards		Topics and Objectives	Days
K.MD.1 K.MD.2	A	**Comparison of Length and Height** Lesson 1: Compare lengths using *taller than* and *shorter than* with aligned and non-aligned endpoints. Lesson 2: Compare length measurements with string. Lesson 3: Make a series of *longer than* and *shorter than* comparisons.	3
K.MD.1 K.MD.2 K.CC.4c K.CC.5 K.CC.6	B	**Comparison of Length and Height of Linking Cube Sticks Within 10** Lesson 4: Compare the length of linking cube sticks to a 5-stick. Lesson 5: Determine which linking cube stick is *longer than* or *shorter than* the other. Lesson 6: Compare the length of linking cube sticks to various objects. Lesson 7: Compare objects using *the same as*.	4
K.MD.1 K.MD.2	C	**Comparison of Weight** Lesson 8: Compare using *heavier than* and *lighter than* with classroom objects. Lesson 9: Compare objects using *heavier than, lighter than*, and *the same as* with balance scales. Lesson 10: Compare the weight of an object to a set of unit weights on a balance scale. Lesson 11: Observe conservation of weight on the balance scale. Lesson 12: Compare the weight of an object with sets of different objects on a balance scale.	5
K.MD.1 K.MD.2	D	**Comparison of Volume** Lesson 13: Compare volume using *more than, less than*, and *the same as* by pouring. Lesson 14: Explore conservation of volume by pouring. Lesson 15: Compare using *the same as* with units.	3

Module 3: Comparison of Length, Weight, Capacity, and Numbers to 10

A STORY OF UNITS

Module Overview K•3

Standards		Topics and Objectives	Days
		Mid-Module Assessment: Topics A–D (Interview style assessment: 3 days)	3
K.CC.6	E	**Are There Enough?** Lesson 16: Make informal comparison of area. Lesson 17: Compare to find if there are enough. Lesson 18: Compare using *more than* and *the same as*. Lesson 19: Compare using *fewer than* and *the same as*.	4
K.CC.6 K.CC.7 K.CC.4c K.MD.2	F	**Comparison of Sets Within 10** Lesson 20: Relate *more* and *less* to length. Lesson 21: Compare sets informally using *more, less*, and *fewer*. Lesson 22: Identify and create a set that has the same number of objects. Lesson 23: Reason to identify and make a set that has 1 more. Lesson 24: Reason to identify and make a set that has 1 less.	5
K.CC.6 K.CC.7 K.CC.4c	G	**Comparison of Numerals** Lesson 25: Match and count to compare a number of objects. State which quantity is more. Lesson 26: Match and count to compare two sets of objects. State which quantity is less. Lesson 27: Strategize to compare two sets. Lesson 28: Visualize quantities to compare two numerals.	4
K.MD.1 K.MD.2 K.CC.6 K.CC.7	H	**Clarification of Measurable Attributes** Lesson 29: Observe cups of colored water of equal volume poured into a variety of container shapes. Lesson 30: Use balls of clay of equal weights to make sculptures. Lesson 31: Use benchmarks to create and compare rectangles of different lengths to make a city. Lesson 32: Culminating task—describe measurable attributes of single objects.	4
		End-of-Module Assessment: Topics E–H (Interview style assessment: 3 days)	3
Total Number of Instructional Days			38

Module 3: Comparison of Length, Weight, Capacity, and Numbers to 10

7

Terminology

New or Recently Introduced Terms

- Balance scale (tool for weight measurement)
- Capacity (with reference to volume)
- Compare (specifically using direct comparison)
- Endpoint (with reference to alignment for direct comparison)
- Enough/not enough (comparative term)
- Heavier than/lighter than (weight comparison)
- Height (vertical distance measurement from bottom to top)
- Length (distance measurement from end to end; in a rectangular shape, length can be used to describe any of the four sides)
- Longer than/shorter than (length comparison)
- More than/fewer than (discrete quantity comparison)
- More than/less than (volume, area, and number comparisons)
- Taller than/shorter than (height comparison)
- The same as (comparative term)
- Weight (heaviness measurement)

Familiar Terms and Symbols[2]

- Match (group items that are the same or that have the same given attribute)
- Numbers 1–10

Suggested Tools and Representations

- Balance scales (as pictured to the right)
- Centimeter cubes
- Clay
- Linking cubes in sticks with a color change at the five
- Plastic cups and containers for measuring volume

---

[2]These are terms and symbols students have seen previously.

Homework

Homework at the K–1 level is not a convention in all schools. In this curriculum, homework is an opportunity for additional practice of the content from the day's lesson. The teacher is encouraged, with the support of parents, administrators, and colleagues, to discern the appropriate use of homework for his or her students. Fluency exercises can also be considered as an alternative homework assignment.

Scaffolds[3]

The scaffolds integrated into *A Story of Units*® give alternatives for how students access information as well as express and demonstrate their learning. Strategically placed margin notes are provided within each lesson elaborating on the use of specific scaffolds at applicable times. They address many needs presented by English language learners, students with disabilities, students performing above grade level, and students performing below grade level. Many of the suggestions are organized by Universal Design for Learning (UDL) principles and are applicable to more than one population. To read more about the approach to differentiated instruction in *A Story of Units*, please refer to "How to Implement *A Story of Units*."

Assessment Summary

Type	Administered	Format	Standards Addressed
Mid-Module Assessment Task	After Topic D	Constructed response with rubric	K.MD.1 K.MD.2
End-of-Module Assessment Task	After Topic H	Constructed response with rubric	K.CC.6 K.CC.7 K.MD.1 K.MD.2
Culminating Task	Lesson 32	Determining the attribute to be measured	K.MD.1 K.MD.2

---

[3]Students with disabilities may require Braille, large-print, audio, or special digital files. Please visit the website www.p12.nysed.gov/specialed/aim for specific information on how to obtain student materials that satisfy the National Instructional Materials Accessibility Standard (NIMAS) format.

Module 3: Comparison of Length, Weight, Capacity, and Numbers to 10

© 2015 Great Minds. eureka-math.org
GK-M3-TE-B3-1.3.1-01.2016

A STORY OF UNITS

Mathematics Curriculum

GRADE K • MODULE 3

Topic A
Comparison of Length and Height

K.MD.1, K.MD.2

Focus Standards:	K.MD.1	Describe measurable attributes of objects, such as length or weight. Describe several measurable attributes of a single object.
	K.MD.2	Directly compare two objects with a measurable attribute in common, to see which object has "more of"/"less of" the attribute, and describe the difference. *For example, directly compare the heights of two children and describe one child as taller/shorter.*
Instructional Days:	3	
Coherence	-Links from:	GPK–M4	Comparison of Length, Weight, Capacity, and Numbers to 5
	-Links to:	G1–M3	Ordering and Comparing Length Measurements as Numbers

In Module 2, students observed, analyzed, and categorized geometric shapes by focusing on their attributes; they now launch into the process of recognizing and comparing these attributes. In Module 3, comparisons of length, weight, and volume transition into comparisons of numbers: longer *than*, shorter *than*, as long as; heavier *than*, lighter *than*, as heavy as; more than, less than, the same as. For example, "8 is *more than* 5. 5 is *less than* 8. 5 is *the same as* 5."

In Topic A, students begin by identifying the attribute of length by determining that a book and a ribbon can be compared in different ways: as longer than, heavier than, or taking up more space. This occurs within the natural context of the lesson, which then proceeds to comparing length and height when endpoints are aligned and not aligned. Jan is shorter than Pat when they are standing next to each other with one of their endpoints automatically aligned. But, what if Jan is standing on a stepladder? Now, the endpoints are not aligned, and students, faced with this complexity, understand that Jan is still shorter than Pat though her head may be higher because she is standing on a stepladder.

In Lesson 2, students compare the length of their strings to the length of various objects within the classroom. "My string is longer than the marker." "My string is shorter than my friend's shoe." They know to line up the endpoints or the comparison is not valid.

In Lesson 3, students make a series of comparisons: the pencil is longer than the marker; the eraser is shorter than the marker. They directly compare only two objects but in doing so, potentially see more relationships. Then, they engage in drawing a magical world where, for example, a flower is taller than a house.

A STORY OF UNITS　　　　　　　　　　　　　　　　　　　　　　　　　　　Topic A　K•3

A Teaching Sequence Toward Mastery of Comparison of Length and Height

Objective 1: Compare lengths using *taller than* and *shorter than* with aligned and non-aligned endpoints.
(Lesson 1)

Objective 2: Compare length measurements with string.
(Lesson 2)

Objective 3: Make a series of *longer than* and *shorter than* comparisons.
(Lesson 3)

Topic A: Comparison of Length and Height

© 2015 Great Minds. eureka-math.org
GK-M3-TE-B3-1.3.1-01.2016

A STORY OF UNITS   Lesson 1   K•3

Lesson 1

Objective: Compare lengths using *taller than* and *shorter than* with aligned and non-aligned endpoints.

Suggested Lesson Structure

- ■ Fluency Practice   (10 minutes)
- ■ Application Problem   (5 minutes)
- ■ Concept Development   (29 minutes)
- ■ Student Debrief   (6 minutes)
- **Total Time**   **(50 minutes)**

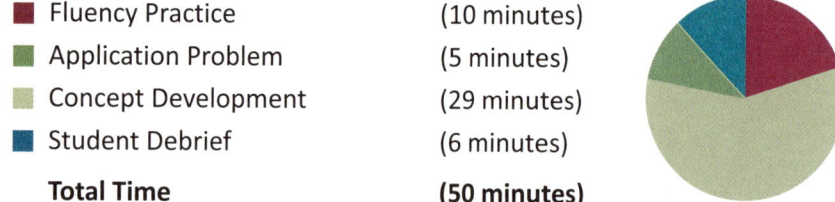

Fluency Practice (10 minutes)

- Tell the Hidden Number **K.CC.2**   (4 minutes)
- 5-Group Finger Counting **K.CC.2**   (2 minutes)
- Say Ten Push-Ups **K.NBT.1**   (4 minutes)

Tell the Hidden Number (4 minutes)

Materials: (S) Pennies, number path (Fluency Template)

Note: This activity maintains students' proficiency in number order, especially starting from a number other than 1. Challenge them by folding the number path to show short sequences (e.g., 4, 5, 6, 7), and have them hide one or two numbers.

Partner A closes his eyes. Partner B hides one of the numbers on the number path with a penny and then tells Partner A to open his eyes. Partner A tells the hidden number. Partners switch roles and play again. Circulate and provide support to students who must count from 1 to determine the hidden number each time.

Variation: Cover two or three numbers with pennies.

5-Group Finger Counting (2 minutes)

Note: This activity helps solidify students' understanding of numbers to 10 in relationship to the five, which will be useful in upcoming lessons.

- T: Quick! Show me 5.
- S: (Extend an open left hand to show 5 without having to count.)
- T: Show me 1 more.
- S: (Show an open left hand for 5 and the thumb of the right hand for 6.)

T: We can count from 5 like this: 5 (push out the left hand), 1 more (push out the thumb of the right hand) is… (push both the left hand and the thumb of the right hand) 6. Try it with me. Ready?

S: 5 (push out the left hand), 1 more (push out the thumb of the right hand) is… (push both the left hand and the thumb of the right hand) 6.

T: Stay there at 6. Now, show me 1 more.

S: (Show an open left hand for 5 and the thumb and the index finger of the right hand for 7.)

T: How many fingers are you showing on your left hand?

S: 5.

T: And your right hand?

S: 2.

T: How many fingers are you showing in all?

S: 7.

T: So, this time, we'll say 5 (push out the left hand), 2 more (push out the thumb and index finger of the right hand) is… (push out both the left hand and the thumb and index finger of the right hand) 7. Try it with me. Ready?

S: 5 (push out the left hand), 2 more (push out the thumb and index finger of the right hand) is… (push out both the left hand and the thumb and index finger of the right hand) 7.

Continue to 10 if students are ready, but there is no need to rush—this is a challenging counting activity. As students begin to note the pattern, steadily remove the scaffold until they can state the relationship to the 5-group without guidance. It would be better for students to achieve mastery to 7 than to mimic the teacher to 10.

Say Ten Push-Ups (4 minutes)

Note: This activity extends students' understanding of numbers to 10 in anticipation of working with teen numbers. Some students may already know how to say the numbers the conventional way. Do not discourage them from making that connection, but perhaps encourage them to say the numbers conventionally in their mind so as to not confuse others.

T: You've gotten so good at counting to ten. It's time to start counting higher! Next is ten 1. Repeat, please.

S: Ten 1.

T: We can show it on our hands like this: ten (push out both hands, palms out, as if doing a push-up exercise in the air, and then pause with closed fists close to body) 1 (push out the right hand pinky finger). It's your turn. Ready?

S: Ten (push out both hands as if doing a push-up exercise in the air) and (closed fists, close to body) 1 (push out the left hand pinky finger).

T: Very good. Next is ten (push out both hands as if doing a push-up exercise in the air) and (closed fists, close to body) 2 (push out the right hand pinky and ring fingers). It's your turn. Ready?

S: Ten (push out both hands as if doing a push-up exercise in the air) and (closed fists, close to body) 2 (push out the left hand pinky and ring fingers).

Ten

and

4

Lesson 1: Compare lengths using *taller than* and *shorter than* with aligned and non-aligned endpoints.

T: Ten (push out both hands as if doing a push-up exercise in the air) and (closed fists, close to body) 3 (push out the right hand pinky, ring, and middle fingers). It's your turn. Ready?

Continue a few more times or until students can count and show the number on their hands fluently from ten to ten 3. In the next lesson, this activity will be extended to ten 5. Consider continuing to ten 5 now if students are ready, as they may catch on to the pattern quickly.

Application Problem (5 minutes)

Materials: (T) Indicated photos, heavy book, piece of ribbon 1 meter long

Setup: Show students a set of photos, one of a skyscraper contrasted with a one-story building.

T: With your partner, look at the photos of the buildings. Talk about how they are the same and how they are different. What do you notice?
S: One is bigger than the other.
T: When you compare and say it is bigger, let's think about what you mean. (After each question, allow students to have a lively, brief discussion.)
T: Do you mean that it is heavier, like this book is heavier than this ribbon? (Dramatize the weight of the book and ribbon.)
T: Do you mean that it is longer, like this ribbon is longer than this book? (Dramatize the length of the ribbon.)
T: Do you mean it takes up more space, like this book takes up more space than this ribbon when it is all squished together? (Dramatize the volume of the book and ribbon.)
T: Do you mean to compare the number of things, like the number of books and ribbons? (Dramatize a count.)
T: So, we can compare things in different ways! Today, let's compare by thinking about how much longer or shorter one thing is than another thing. (Dramatize.)

> **NOTES ON MULTIPLE MEANS OF ACTION AND EXPRESSION:**
>
> The teacher can help students who are below grade level practice the differences between the concepts of *taller than* and *shorter than* by using interactive technology, such as the game found at the website:
> http://www.softschools.com/measurement/games/tall_and_short/

Concept Development (29 minutes)

Materials: (T) 2 chairs, 2 different lengths of string, 2 pencils of different lengths (S) 2 strips of paper (a longer blue one and a shorter red one)

T: We are going to have a magic show! Student A, please come stand by me. Class, what do you notice about our heights?
S: You are bigger.

Lesson 1: Compare lengths using *taller than* and *shorter than* with aligned and non-aligned endpoints.

A STORY OF UNITS   **Lesson 1** K•3

T: Yes, I am **taller than** Student A. We say that Student A is **shorter than** I am. Now, watch my magic. Abracadabra! (Pull out two chairs. Sit on one chair, and ask Student A to stand on the other.)

T: It's magic! Isn't Student A taller than I am now?

S: No. She's standing on the chair!

T: So, even though her head is above mine right now, am I still taller than Student A?

S: Yes!

T: Hmmm. Thanks anyway, Student A.

T: Student B, could you please come help me? (Hand Student B two pieces of string of differing lengths, and have the student hold them up for class observation.)

T: Student B, what do you notice about the strings?

S: This one is longer!

T: This string is **longer than** that one. Abracadabra! (Take longer string, and fold it several times to make it shorter than the other. Hand it back to the student.) Now, it is shorter than the other. It's magic!

S: No, it isn't! You just crumpled it up, but it is still longer.

T: Oh well. Thank you, Student B.

T: I have two pencils. (Show students pencils of differing lengths.) This pencil is shorter than the other one. Now, close your eyes. (Place the pencils in your fist so that they appear to be equal.) Abracadabra!

T: Look at the pencils now. They are the same length! It's magic! (Varied responses.)

T: Student C, come look at my pencils, and tell the class what you see. (Have Student C observe the pencils.)

S: They aren't the same. → You were hiding the bottoms! The bottoms have to be even. → This one is really longer.

T: You are right. The **endpoints** of the pencils need to be in the same place for us to **compare** them fairly. Now, you will get a chance to be the magicians. You and your partner will have two strips of paper. Compare to see which one is longer.

S: The blue one.

T: With your partner, see if you can find a way to make the red one look longer than the blue one. (Allow students time to arrange the strips and experiment.) What happens if you line up the ends of the strips with the edge of your desk?

MP.6

S: Now, they start in the same place. → We can see that the blue one really is longer! → That is the fair way to do it.

T: This reminds me of the number work we did with counters. Remember, even when we moved our counters around in different ways, we still had the same number of things. How is that similar to what you just saw?

S: Even if you move things around, they are still just as tall as they were before. (Guide students to realize that the attribute of length is conserved regardless of orientation or endpoint alignment. Encourage them to articulate the necessity of accurate alignment.)

T: Now, we will think about taller than and shorter than while we look at our Problem Set.

Lesson 1: Compare lengths using *taller than* and *shorter than* with aligned and non-aligned endpoints.

© 2015 Great Minds. eureka-math.org
GK-M3-TE-B3-1.3.1-01.2016

15

A STORY OF UNITS

Lesson 1 K•3

Problem Set (10 minutes)

Students should do their personal best to complete the Problem Set within the allotted time.

For some classes, it may be appropriate to modify the assignment by specifying which problems students should work on first. With this option, let the purposeful sequencing of the Problem Set guide your selections so that problems continue to be scaffolded. Balance word problems with other problem types to ensure a range of practice. Consider assigning incomplete problems for homework or at another time during the day.

Student Debrief (6 minutes)

Lesson Objective: Compare lengths using *taller than* and *shorter than* with aligned and non-aligned endpoints.

The Student Debrief is intended to invite reflection and active processing of the total lesson experience.

Invite students to review their solutions for the Problem Set. They should check work by comparing answers with a partner before going over answers as a class. Look for misconceptions or misunderstandings that can be addressed in the Debrief. Guide students in a conversation to debrief the Problem Set and process the lesson.

Any combination of the questions below may be used to lead the discussion.

- What did you notice when we were looking at the pencils? (Note: Guide discussion to focus on the importance of proper **endpoint** alignment.)
- What did you notice when we were **comparing** the strings? (Note: Guide discussion to focus on conservation of length with respect to orientation and movement.)
- How did you know which paper strip on your Problem Set was **longer than** the other?
- How did you know which paper strip on your Problem Set was **shorter than** the other?
- Explain to your partner how you were able to draw the flower **taller than** the vase. Did your partner think the same way?

Lesson 1: Compare lengths using *taller than* and *shorter than* with aligned and non-aligned endpoints.

- When we started our lesson, we thought about how we might compare things. What were we comparing today? How heavy something is, how long something is, how many of something there are, or how much space something takes up?

Homework

Homework at the K–1 level is not a convention in all schools. In this curriculum, homework is an opportunity for additional practice of the content from the day's lesson. The teacher is encouraged, with the support of parents, administrators, and colleagues, to discern the appropriate use of homework for his or her students. Fluency exercises can also be considered as an alternative homework assignment.

Lesson 1: Compare lengths using *taller than* and *shorter than* with aligned and non-aligned endpoints.

A STORY OF UNITS

Lesson 1 Problem Set K•3

Name _____ Date _____

For each pair, circle the longer one. Imagine the paper strips are lying flat on a table.

Draw a flower that is taller than the vase.

Draw a tree that is taller than the house.

For each pair, circle the shorter one.

Draw a bookmark that is shorter than this book.

Draw a crayon that is shorter than this pencil.

A STORY OF UNITS Lesson 1 Homework K•3

Name _____ Date _____

Draw 3 more flowers that are shorter than these flowers. Count how many flowers you have now. Write the number in the box.

☐

Draw 2 more ladybugs that are taller than these ladybugs. Count how many ladybugs you have now. Write the number in the box.

☐

On the back of your paper, draw something that is taller than you. Draw something that is shorter than a flagpole.

Lesson 1: Compare lengths using *taller than* and *shorter than* with aligned and non-aligned endpoints.

| 1 | 2 | 3 | 4 | 5 | 6 | 7 | 8 | 9 | 10 |

| 1 | 2 | 3 | 4 | 5 | 6 | 7 | 8 | 9 | 10 |

| 1 | 2 | 3 | 4 | 5 | 6 | 7 | 8 | 9 | 10 |

| 1 | 2 | 3 | 4 | 5 | 6 | 7 | 8 | 9 | 10 |

| 1 | 2 | 3 | 4 | 5 | 6 | 7 | 8 | 9 | 10 |

| 1 | 2 | 3 | 4 | 5 | 6 | 7 | 8 | 9 | 10 |

number path

Lesson 1: Compare lengths using *taller than* and *shorter than* with aligned and non-aligned endpoints.

A STORY OF UNITS Lesson 2 K•3

Lesson 2

Objective: Compare length measurements with string.

Suggested Lesson Structure

- ■ Fluency Practice   (10 minutes)
- ■ Application Problem (5 minutes)
- ■ Concept Development (29 minutes)
- ■ Student Debrief   (6 minutes)
- **Total Time**   **(50 minutes)**

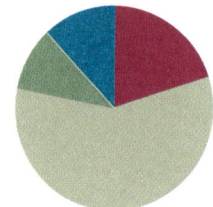

Fluency Practice (10 minutes)

- Show Me Taller and Shorter K.MD.1   (3 minutes)
- Say Ten Push-Ups K.NBT.1   (3 minutes)
- Make It Equal K.CC.6   (4 minutes)

Show Me Taller and Shorter (3 minutes)

Materials: (T) Marker, crayon

Note: This activity recalls the previous lesson's work with height, connecting to today's work with length.
- T: Let's use our hands to show taller and shorter. For taller, we'll do this (hold one hand above your head and the other at waist level, indicating height). Show me taller.
- S: (Show the hand gesture for taller.)
- T: To show shorter, we'll do this (hold hands closer than before, indicating a shorter height).
- S: (Show the gesture for shorter.)
- T: Let's practice. Show me taller.
- S: (Show the hand gesture for taller.)
- T: Show me shorter.
- S: (Show the gesture for shorter.)

Mix it up, and quicken the pace to see that students understand the meaning of the vocabulary.
- T: Look at my marker (hold a marker upright), and look at my crayon. Is the crayon shorter or taller?
- T: Show me the gesture for taller if you think the crayon is taller. Show me the gesture for shorter if you think the crayon is shorter.
- S: (Demonstrate either shorter or taller gesture.)

Use a couple more items for demonstration of shorter, taller (e.g., book, pencil).

Lesson 2: Compare length measurements with string.

A STORY OF UNITS — Lesson 2 K•3

Say Ten Push-Ups (3 minutes)

Note: This activity extends students' understanding of numbers to 10 in anticipation of working with teen numbers.

Conduct activity as described in Lesson 1, but now continue to ten 5.

Make It Equal (4 minutes)

Materials: (S) Bags of beans, laminated paper or foam work mat, dice

Note: Students develop a visual sense of comparison in this activity, preparing them to compare lengths of objects in this lesson.

1. Teacher introduces the term *equal* as meaning *the same number*.
2. Both partners roll the dice and put the same number of beans on their work mat as dots shown on the dice.
3. Partner A has to make his beans equal to his partner's by taking off or putting on more beans.
4. Partner B counts to verify.
5. Switch roles and play again.

Application Problem (5 minutes)

Draw a picture of something you have seen that is very tall. Compare your picture to your friend's. Is the item in her drawing taller than or shorter than yours? Are you sure? How can you find out?

Note: This application problem serves as a review of the vocabulary of yesterday's lesson and allows the students to practice proper endpoint alignment in comparison. They use this skill in today's lesson. Circulate to ensure they are comparing accurately.

> **NOTES ON MULTIPLE MEANS OF REPRESENTATION:**
>
> Highlight the critical vocabulary of *taller than* and *shorter than* for English language learners by showing a visual of the words as they are taught. This helps them follow the lesson and engage with the key concepts of the lesson. Include the words *taller than* and *shorter than* with the visual on the word wall after the lesson.

Lesson 2: Compare length measurements with string.

A STORY OF UNITS   Lesson 2   K•3

Concept Development (29 minutes)

Materials: (T) String, scissors, masking tape (S) String, scissors, clipboard, pencil, longer or shorter recording sheet (Template)

Note: Have students save their string, as they need it for their homework.

T: Today, your job will be to compare the **length** or **height** of things in our classroom to the length of a piece of string. You will each have a string of your own to use. First, I will make one for myself. (Cut a piece of string approximately one foot long, and show it to the students. Label the string with a piece of masking tape and your initials.)

T: I want my string to be this long. Now, I want to compare it to some things in the room. Let's make a chart. (On the board, create a quick chart as follows.)

These things are *longer than* my string.	These things are *shorter than* my string.

T: I'm going to test a few things to show how this works. Look at my desk. (Review and model correct endpoint alignment, lining the string up on the edge of the desk.) Is my desk longer than or shorter than my string?

S: Longer.

T: Can you say, "Longer than your string"?

S: Longer than your string.

T: Let's draw that on the chart. (Repeat with a few other examples. Model correct technique until students understand how to make precise comparisons.)

T: Now, you and your partner will help each other make your special measuring strings. Show your partner how long you would like your string to be, and then he can help you cut it. Be sure to label your strings with a piece of masking tape and your initials, because otherwise, they will look a lot alike! (Assist as necessary while students prepare their measuring tools. While it is not necessary that all of the strings be the same, encourage students to use reasonable lengths.)

T: Here are clipboards and your own charts just like the one on the board. Use your strings to compare lengths. Find at least five things that are longer than your string and at least five things that are shorter than your string. Draw them on your charts. Maybe you will discover something that is the same length as your string! If you do, draw it on the back of your sheet. (Allow time for exploration, measurement, and recording.)

T: Who would like to show and share some things that he or she discovered? Did you find any things that are almost the same length as your string? (Allow a few minutes for discussion.)

T: Put your string in your pocket or backpack. You can measure more things after school and at home tonight.

24   Lesson 2:   Compare length measurements with string.

A STORY OF UNITS — Lesson 2 K•3

Problem Set (10 minutes)

Students should do their personal best to complete the Problem Set within the allotted time.

On this Problem Set, have students compare as many pictures as they can. For the sake of time, students could circle or just put a line of color on the object for longer or shorter to stay within the timeframe.

Student Debrief (6 minutes)

Lesson Objective: Compare length measurements with string.

The Student Debrief is intended to invite reflection and active processing of the total lesson experience.

Invite students to review their solutions for the Problem Set. They should check work by comparing answers with a partner before going over answers as a class. Look for misconceptions or misunderstandings that can be addressed in the Debrief. Guide students in a conversation to debrief the Problem Set and process the lesson.

Any combination of the questions below may be used to lead the discussion.

- What did you notice as you compared each object to the string?
- Did you do anything different as you compared the **lengths**? What did you need to be sure to do? Why?
- Did you predict if the string would be shorter than or longer than before you measured?
- Explain to your partner how you compared the **heights**. Did your partner do anything different?
- Does it matter which way you compare two objects? Why? How did you compare the string and the door?
- Did your partner find something that was longer for his string that was shorter for yours? Did she find something that was shorter for her string that was longer for yours? Why did that happen?
- What new math vocabulary did we use today to communicate precisely?
- How did the Application Problem connect to today's lesson?

Lesson 2: Compare length measurements with string.

Name _____ Date _____

Cut out the picture of the string at the bottom of the page. Compare the string with each object to see which is longer. Use the line next to each object to help you compare. Color objects shorter than the string green. Color objects longer than the string orange.

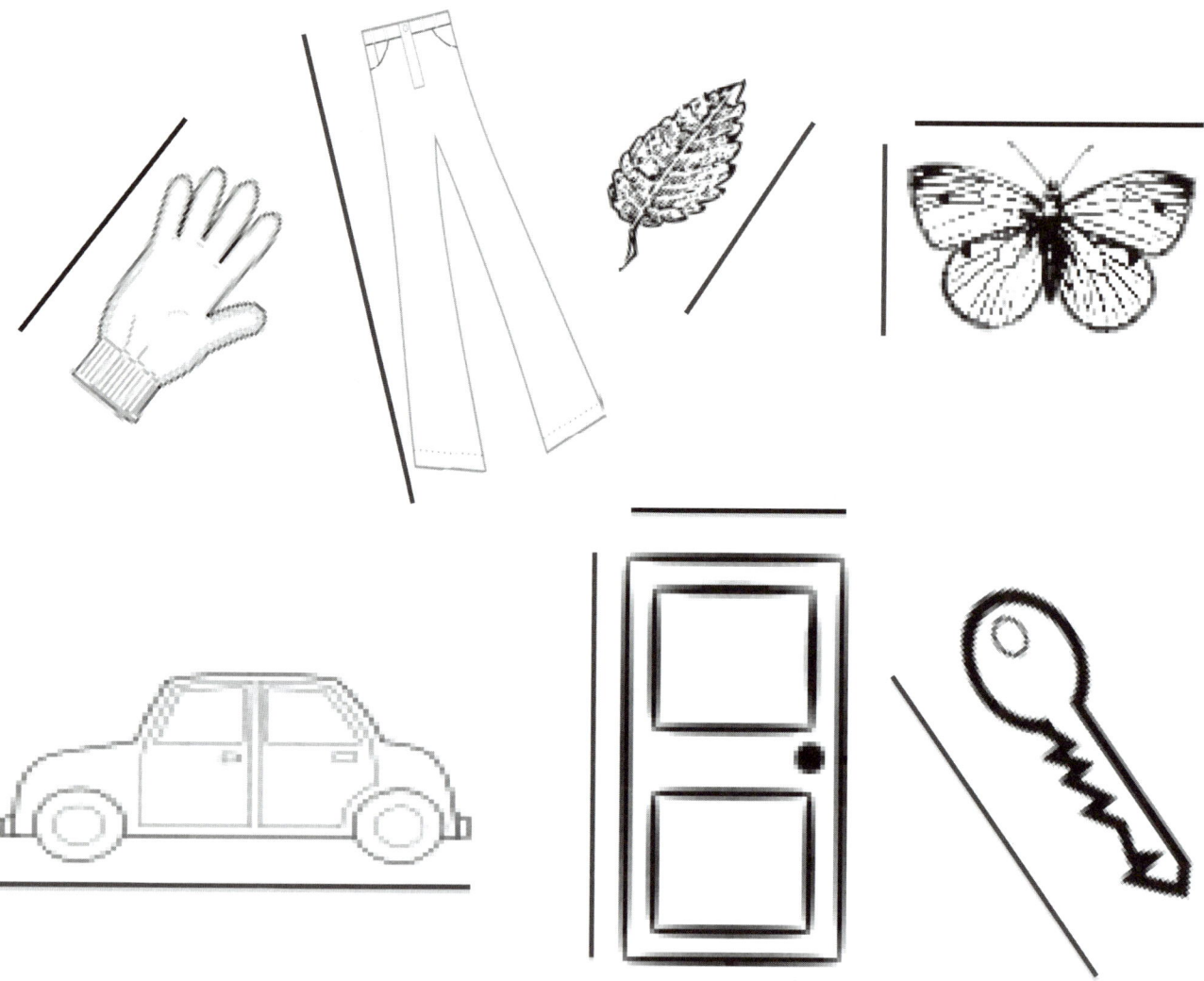

On the back of your paper, draw something longer than, shorter than, and the same length as the picture of the string. Color objects shorter than the string green. Color objects longer than the string orange.

A STORY OF UNITS   Lesson 2 Homework K•3

Name _____   Date _____

Using the piece of string from class, find three items at home that are shorter than your piece of string and three items that are longer than your piece of string. Draw a picture of those objects on the chart. Try to find at least one thing that is about the same length as your string, and draw a picture of it on the back.

Shorter than the string	Longer than the string

Lesson 2:   Compare length measurements with string.

Longer or Shorter Recording Sheet

These objects are **longer than** my string.	These objects are **shorter than** my string.

_____
longer or shorter

A STORY OF UNITS

Lesson 3 K•3

Lesson 3

Objective: Make a series of *longer than* and *shorter than* comparisons.

Suggested Lesson Structure

   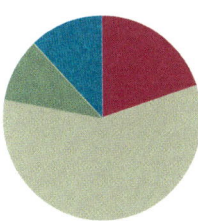

Fluency Practice (10 minutes)

- Say Ten Push-Ups **K.NBT.1**   (3 minutes)
- Hidden Numbers (5 as the Whole) **K.OA.3**   (4 minutes)
- Make It Equal **K.CC.6**   (3 minutes)

Say Ten Push-Ups (3 minutes)

Conduct activity as outlined in Lesson 1, but now continue to ten 5, encouraging students to predict what comes next in the pattern.

Note: This activity extends students' understanding of numbers to 10 in anticipation of working with teen numbers.

Hidden Numbers (5 as the Whole) (4 minutes)

Materials: (S) Hidden numbers mat (Fluency Template) inserted into personal white board

Note: Finding embedded numbers anticipates the work of Kindergarten Module 4 by developing part–whole thinking.

    T: Touch and count the fish on your mat. Raise your hand when you know how many. (Wait for all hands to go up, and then give the signal.) Ready?

    S: 10.

    T: Put an X on 5 of the fish. We're not going to count those fish right now. Pretend they swam away!

    S: (Cross out 5 fish.)

    T: Circle a group of 4 from the fish who didn't swim away.

    T: How many fish are left?

    S: 1.

Lesson 3: Make a series of *longer than* and *shorter than* comparisons.

29

A STORY OF UNITS   Lesson 3   K•3

T: Let's circle that 1. How many did you circle all together?
S: 5.
T: Erase your board. Put an X on 5 of the fish again to show they swam away. How many fish did not swim away?
S: 5.
T: Now, this time, circle a group of 2. Circle another 2.
S: (Circle two groups of 2.)
T: How many fish have you circled so far?
S: 4.
T: Circle 1 more. Now, how many are circled?
S: 5.
T: Erase your boards. Put an X on 5 of the fish again to show they swam away. How many fish did not swim away?
S: 5.
T: This time, circle a group of 3.
T: Circle a group of 2.
T: How many are in the larger group?
S: 3.
T: How many are in the smaller group?
S: 2.
T: How many did you circle all together?
S: 5.

Continue this procedure, looking for hidden numbers within a group of 6. Pause occasionally to allow students to explain efficient ways of locating the groups.

Make It Equal (3 minutes)

Conduct activity as outlined in Lesson 2, but now have students line up their beans (up to 10 beans) in horizontal rows or vertical columns.

Note: In this activity, students experience comparison visually, a skill crucial to the work of this module.

NOTES ON MULTIPLE MEANS OF REPRESENTATION:

Students working below grade level benefit from extra practice in determining what objects are longer than or shorter than. This helps prepare them for comparing two different lengths with a third object in this lesson. Use interactive technology, such as the following website:
http://www.softschools.com/measurement/games/long_and_short/.

Application Problem (5 minutes)

Draw a monkey with a very long tail. Draw a monkey with a very short tail. Now, draw a yummy banana for the monkeys to share. Is the banana longer than or shorter than the tail of the first monkey? Is it longer than or shorter than the tail of the second monkey? Tell your partner what you notice.

Note: The comparison of two different lengths with a neutral object introduces today's lesson objective.

30   Lesson 3:   Make a series of *longer than* and *shorter than* comparisons.

© 2015 Great Minds. eureka-math.org
GK-M3-TE-B3-1.3.1-01.2016

A STORY OF UNITS   Lesson 3   K•3

Concept Development (29 minutes)

Materials: (S) Longer than and shorter than work mat (Template), popsicle stick and prepared paper bag filled with various items to measure (e.g., pencil, eraser, glue stick, toy car, small block, 12-inch piece of string, marker, child's scissors, crayon, tower of 5 linking cubes) per pair

MP.6

T: Today, you and your partner have a mystery bag! Each of you close your eyes, and take something out of the bag. Put the objects on your desk.

T: Here is a popsicle stick. Take one of your objects, and compare its length to the popsicle stick. (Select a pair of students to demonstrate. Model and have students repeat. Correct *longer than* and *shorter than* language, if necessary.) Student A, what do you notice?

S: This car is shorter than the popsicle stick.

T: Student B?

S: This pencil is longer than the popsicle stick.

T: Take out another object, and compare it to the popsicle stick. Tell your partner what you observe. (Allow time for students to compare the rest of the objects in the bag with the stick.)

T: How could we use the popsicle stick to help us sort these objects?

S: By size! → We could find all of the things that are longer than the length of the stick and the ones that are shorter than the length of the stick.

T: Good idea. Here is a work mat to help you with your sort. (Distribute work mats to students, and allow them to begin. During the activity, students may line up objects by size within the sort category. Acknowledge correct examples of this, but do not require it.)

T: What if you put away your popsicle stick and used your toy car instead to help you sort?

S: The sort would come out differently. → This would have to go on the other side!

T: Which objects would you need to move? Let's find out. This time, use your toy car to measure the other things. (Continue the exercise through several iterations, each time sorting with respect to the length of a different object from the bag.)

T: Did you notice anything during your sorting?

S: It changes every time! → When we used the little eraser to sort, everything was on one side. → When we used the string, everything else was on one side. The string was the longest thing.

T: Put your objects back in the bag. Let's use our imaginations to think about length in a different way as we complete our Problem Set activity.

> **NOTES ON MULTIPLE MEANS OF ACTION AND REPRESENTATION:**
>
> Challenge students working above grade level by extending the task. Ask them, individually or in teams, to order the objects in their mystery bags from shortest to longest. Also, ask them to find objects in the classroom that can be added to everyone's mystery bag.

> **NOTES ON MULTIPLE MEANS OF REPRESENTATION:**
>
> Modify the directions on the Problem Set as necessary depending on the overall ability level of the class. If students seem to tire, curtail the exercise after drawing a few of the objects. If they are adept at the exercise, give some extra time for the extension activity at the end of the story.

Lesson 3: Make a series of *longer than* and *shorter than* comparisons.

A STORY OF UNITS   Lesson 3 K•3

Problem Set (10 minutes)

Students should do their personal best to complete the Problem Set within the allotted time.

Read the directions carefully to the students. Consider using a timer to limit the sketching of each object, leaving a couple of minutes toward the end during which the students may fill in details of their drawing. Circulate during the activity to assess understanding.

Student Debrief (6 minutes)

Lesson Objective: Make a series of *longer than* and *shorter than* comparisons.

The Student Debrief is intended to invite reflection and active processing of the total lesson experience.

Invite students to review their solutions for the Problem Set. They should check work by comparing answers with a partner before going over answers as a class. Look for misconceptions or misunderstandings that can be addressed in the Debrief. Guide students in a conversation to debrief the Problem Set and process the lesson.

Any combination of the questions below may be used to lead the discussion.

- What did you notice when you changed the object you were comparing within our mystery bag activity?
- What did you think about when you were deciding how to draw the pirate's daughter?
- What did you think about when you were deciding how to draw your caterpillar? How were the words *longer than* and *shorter than* useful when you were telling your partner about your picture?

Directions: Pretend that I am a pirate who has traveled far away from home. I miss my house and family. Will you draw a picture as I describe my home? Listen carefully, and draw what you hear.

- Draw a house in the middle of the paper as tall as your pointer finger.

- Now, draw my daughter. She is shorter than the house.

- There's a great tree in my yard. My daughter and I love to climb the tree. The tree is taller than my house.

- My daughter planted a beautiful daisy in the yard. Draw a daisy that is shorter than my daughter.

- Draw a branch lying on the ground in front of the house. Make it the same length as the house.

- Draw a caterpillar next to the branch. My parrot loves to eat caterpillars. Of course, the length of the caterpillar is shorter than the length of the branch.

- My parrot is always hungry, and there are plenty of bugs for him to eat at home. Draw a ladybug above the caterpillar. Should the ladybug be shorter or longer than the branch?

- Now, draw some more things you think my family would enjoy.

Show your picture to your partner, and talk about the extra things that you drew. Use *longer than* and *shorter than* when you are describing them.

Name _____ Date _____

Home is where the heARRt is, matey.

Lesson 3: Make a series of *longer than* and *shorter than* comparisons.

A STORY OF UNITS Lesson 3 Homework K•3

Name _____ Date _____

Take out a new crayon. Circle objects with lengths shorter than the crayon blue. Circle objects with lengths longer than the crayon red.

On the back of your paper, draw some things shorter than the crayon and longer than the crayon. Draw something that is the same length as the crayon.

Lesson 3: Make a series of *longer than* and *shorter than* comparisons.

Longer than...

Shorter than...

longer than and shorter than work mat

hidden numbers mat

Lesson 3: Make a series of *longer than* and *shorter than* comparisons.

A STORY OF UNITS

Mathematics Curriculum

GRADE K • MODULE 3

Topic B

Comparison of Length and Height of Linking Cube Sticks Within 10

K.MD.1, K.MD.2, K.CC.4c, K.CC.5, K.CC.6

Focus Standards:	K.MD.1	Describe measurable attributes of objects, such as length or weight. Describe several measurable attributes of a single object.
	K.MD.2	Directly compare two objects with a measurable attribute in common, to see which object has "more of"/"less of" the attribute, and describe the difference. *For example, directly compare the heights of two children and describe one child as taller/shorter.*
Instructional Days:	4	
Coherence -Links from:	GPK–M4	Comparison of Length, Weight, Capacity, and Numbers to 5
-Links to:	G1–M3	Ordering and Comparing Length Measurements as Numbers

In Topic A, students compared length and height of different objects when their endpoints were aligned and not aligned. Topic B continues with informal comparison of length with students comparing the lengths and heights of linking cube sticks within 10 with a color change at 5. In Lesson 4, to reinforce the importance of the 5-group, students compare multi-unit linking cube sticks to a 5-stick. "My 4-stick is shorter than my 5-stick."

In Lesson 5, students compare lengths with endpoints that are aligned and not aligned. "My 7-stick is longer than my 4-stick. When I push my 4-stick up or turn it on an angle, it is still shorter than my 7-stick."

In Lesson 6, students compare their linking cube sticks to objects. "My 4-stick is shorter than my pencil. My 4-stick is longer than my eraser." Using linking cubes to directly compare different objects is a precursor to being able to compare the lengths of two objects using a third object and order the lengths of different objects in later grades, as well as provide students with a practical context for solidifying their developing number sense.

In Lesson 7, the students break their 5-stick into two parts. "I broke my 5-stick into two parts. My 5-stick is longer than my 3- or 2-sticks. Together, my 3- and 2-sticks are the same as my 5-stick." This is an extension of their decomposition work from Kindergarten Module 1. This provides the foundation for the number work coming in Kindergarten Module 4, wherein students decompose all numbers to 10. This also encourages their fluency with facts to 5.

A STORY OF UNITS   Topic B

A Teaching Sequence Toward Mastery of Comparison of Length and Height of Linking Cube Sticks Within 10

Objective 1: Compare the length of linking cube sticks to a 5-stick.
(Lesson 4)

Objective 2: Determine which linking cube stick is *longer than* or *shorter than* the other.
(Lesson 5)

Objective 3: Compare the length of linking cube sticks to various objects.
(Lesson 6)

Objective 4: Compare objects using *the same as*.
(Lesson 7)

Lesson 4

Objective: Compare the length of linking cube sticks to a 5-stick.

Suggested Lesson Structure

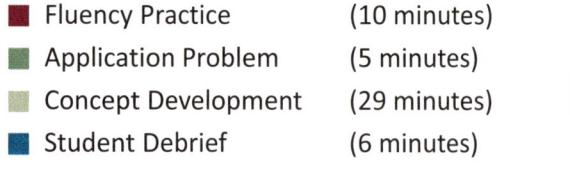

- Fluency Practice   (10 minutes)
- Application Problem (5 minutes)
- Concept Development (29 minutes)
- Student Debrief   (6 minutes)
- **Total Time**   **(50 minutes)**

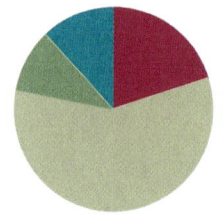

Fluency Practice (10 minutes)

- Show Me Longer and Shorter K.MD.1   (3 minutes)
- Show Me Fingers the Say Ten Way K.NBT.1   (4 minutes)
- 5-Group Finger Counting K.CC.2   (3 minutes)

Show Me Longer and Shorter (3 minutes)

Note: This kinesthetic fluency activity reviews vocabulary.

Conduct the activity as described in Lesson 2, but with *longer* and *shorter*. Now, students extend their hands from side to side to indicate length.

Show Me Fingers the Say Ten Way (4 minutes)

T: You're getting very good at counting on your fingers the Say Ten way! Show me ten 1.
S: Ten (push out both hands as if doing a push-up exercise in the air) and (closed fists, close to body), 1 (push out the left hand pinky finger).
T: Show me ten 2.
S: Ten (push out both hands as if doing a push-up exercise in the air) and (closed fists, close to body), 2 (push out the left hand pinky and ring fingers).

Continue in a predictable pattern and then randomly.

5-Group Finger Counting (3 minutes)

Note: This activity helps solidify students' understanding of numbers to 10 in relationship to the five, which is useful in upcoming lessons.

Conduct the activity as described in Lesson 1.

A STORY OF UNITS Lesson 4 K•3

Application Problem (5 minutes)

Write the following sentence frame on the board, and then read it to the students.

I am taller than _____. I am shorter than _____.

Draw two things on your paper that would make your sentence true. Tell your sentence to your partner. Does he agree that it is true?

Note: Mentally comparing the height or length of two different objects to a third object provides a good cumulative review of the topic to date. This helps students be fully prepared for a more abstract development of measurement in upcoming lessons. Circulate during the activity to ensure that students are completing the sentence.

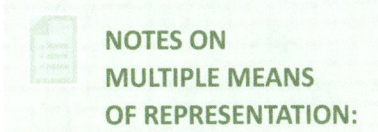

NOTES ON MULTIPLE MEANS OF REPRESENTATION:

Scaffold the Application Problem for English language learners, as well as students with disabilities, if needed, by asking questions such as, "Which is longer, the crayon or the ruler?" and "Which is shorter, the paper clip or the pencil?" Continue to ask questions, gradually leading students to independence.

Concept Development (29 minutes)

Materials: (S) Bag of loose linking cubes per pair: 40 red and 15 of another color or 30 of one color and 25 of another, depending on how you build the stairs (the latter is pictured below), longer or shorter mat (Template)

- T: Do you remember the number stairs we made earlier this year? With your partner, make a set of red number stairs from 1 to 5.
- S: (Create stairs.)
- T: What did we do to make the rest of the stairs?
- S: We made a bunch of 5-sticks, and then we put other cubes on top.
- T: You have great memories! Let's do that again. Use the rest of your red cubes (or orange cubes) to make as many 5-sticks as you can. Then, add your other cubes to make the rest of your number stairs. Put them in order so you make sure you have them all.
- S: (Complete and arrange number stairs.)
- T: What do you notice about the number stairs?
- S: Some are all red. → Some have two colors. → Some are longer. → Some are shorter.
- T: Let's count to make sure we aren't missing any!
- S: 1. 1 more is 2. 2. 1 more is 3. 3. (Continue the pattern through 10.) 10.

Lesson 4: Compare the length of linking cube sticks to a 5-stick. 41

EUREKA MATH

A STORY OF UNITS   Lesson 4   K•3

MP.2

T: Now, mix them all up. Can you find your 5-stick? Hold it up for me to see. How many cubes?

S: 5.

T: Let's compare with your 5-stick! We will use this new work mat to help us organize the rest of the stairs. (Pass out work mat.) Choose another cube stick from your desk. Is that stick longer than or shorter than your 5-stick? (Encourage students to use the sentence, "My _____ stick is shorter than/longer than my _____ stick.")

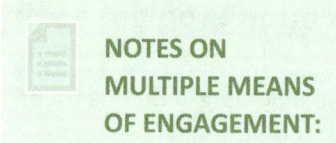

S: (Answers vary.)

T: If your stick is longer than your 5-stick, put it on this side of the mat. (Demonstrate.) If it is shorter, put it on this side. (Demonstrate.) Choose another one. Compare it to your 5-stick. Which side should it go on? (Continue the activity until all sticks have been compared.)

T: Take all of the sticks off your mat, and mix them up again on your desk. Find your 5-stick. This time, I am going to see how long it takes you to measure and sort your sticks onto your work mat. Ready? Set. Go! (Count while students quickly sort sticks. If necessary, repeat the activity until students demonstrate fluency and confidence with comparing and sorting.)

T: Great! Now, take a minute to look at your work mat with your partner. Talk about what you notice about the sticks that you sorted. (Circulate to observe discussions. Observe to determine whether students are making the connection between sticks longer than/shorter than the 5-stick and numbers that are greater than/less than 5. Also, observe to determine whether students detect the connection between length and color.)

> **NOTES ON MULTIPLE MEANS OF ENGAGEMENT:**
>
> Push the comprehension of students working above grade level by asking them to explain and defend their placement of shorter than/longer than linking cube sticks to a friend who is visually impaired.

This would be a good time to call on students to make a comparison using the sentence, "My _____ stick is shorter/longer than my _____ stick."

T: Put your sticks away carefully because we will be using them again tomorrow.

Problem Set (10 minutes)

Students should do their personal best to complete the Problem Set within the allotted time.

Student Debrief (6 minutes)

Lesson Objective: Compare the length of linking cube sticks to a 5-stick.

The Student Debrief is intended to invite reflection and active processing of the total lesson experience.

Invite students to review their solutions for the Problem Set. They should check work by comparing answers with a partner before going over answers as a class. Look for misconceptions or misunderstandings that can be addressed in the Debrief. Guide students in a conversation to debrief the Problem Set and process the lesson.

Any combination of the questions below may be used to lead the discussion.

- How did you compare the sticks in the sorting activity? (Review the importance of endpoint alignment.)
- Was it easier to sort the sticks the second time? Why?
- When you were sorting the sticks, did you notice any patterns?
- Did you notice any clues from the colors of the sticks that helped you with your sort?

Name _____ Date _____

Circle the shorter stick.

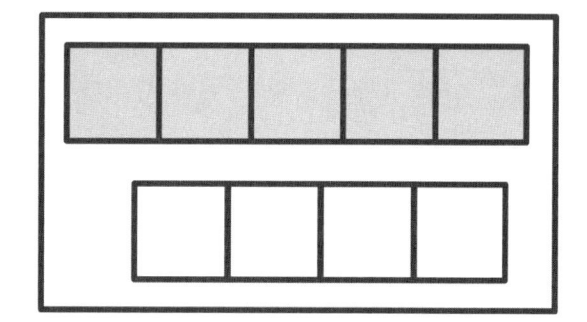

How many linking cubes are in the shorter stick? Write the number in the box.

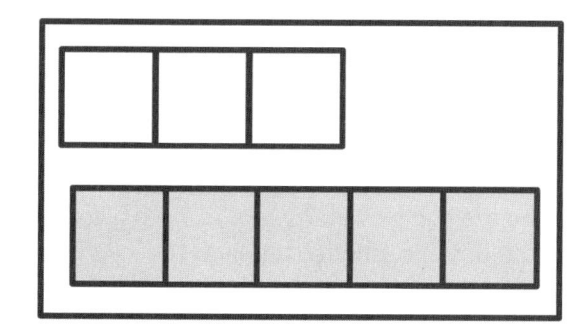

How many linking cubes are in the shorter stick? Write the number in the box.

Circle the longer stick.

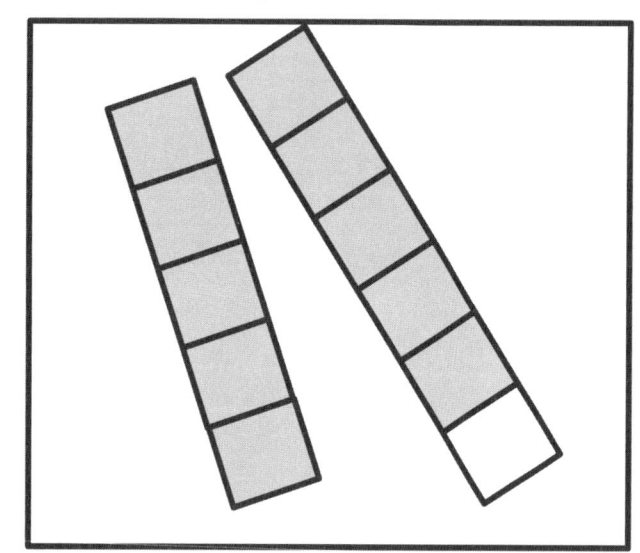

How many linking cubes are in the longer stick? Write the number in the box.

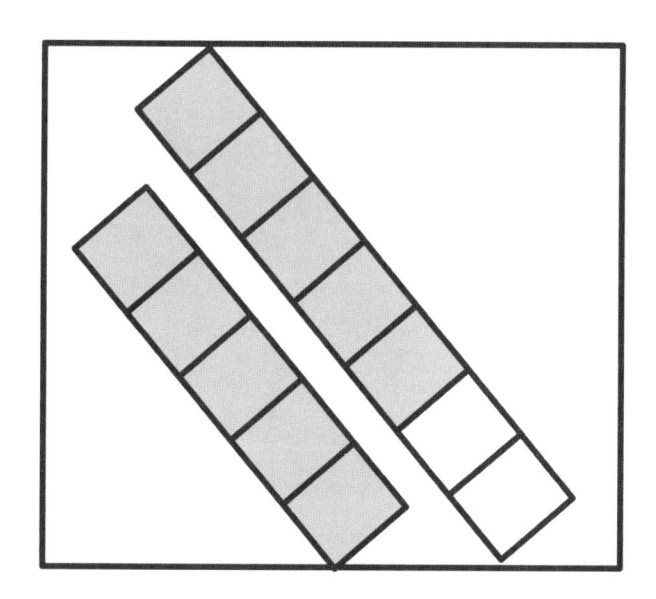

How many linking cubes are in the longer stick? Write the number in the box.

A STORY OF UNITS

Lesson 4 Problem Set K•3

Draw a stick **shorter than** my 5-stick.

Draw a stick **longer than** mine.

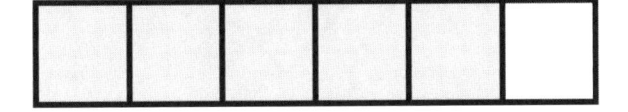

Draw a stick **shorter than** mine.

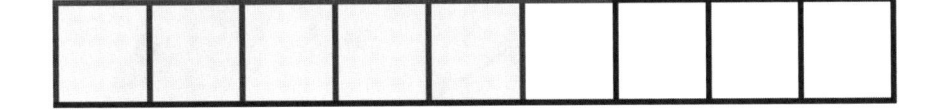

Lesson 4: Compare the length of linking cube sticks to a 5-stick.

Name _____   Date _____

Use a red crayon to circle the sticks that are shorter than the 5-stick.

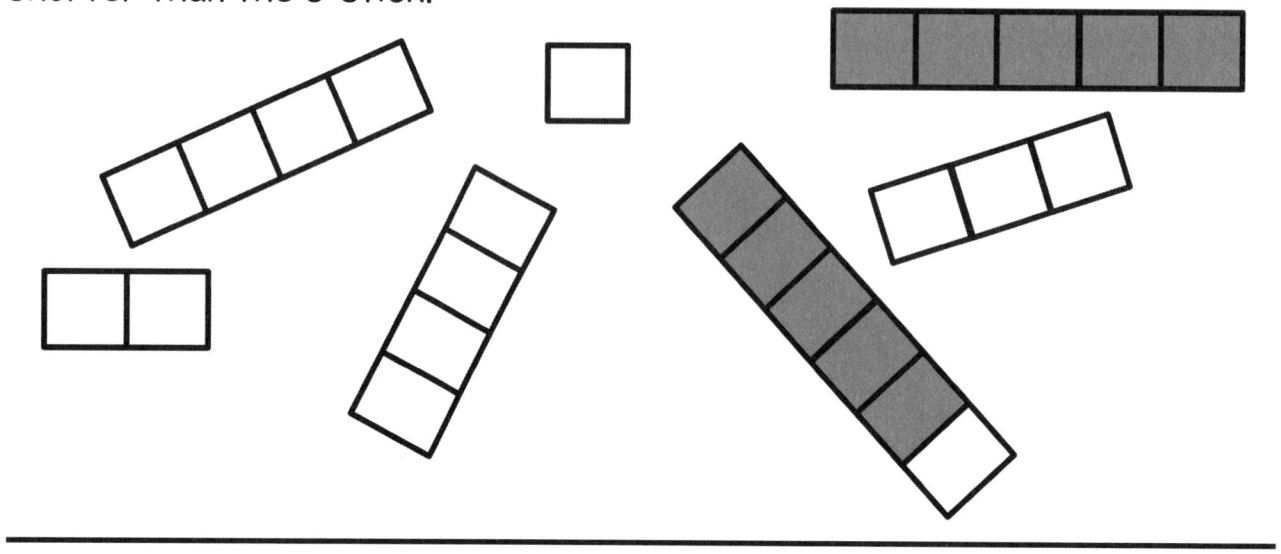

Use a blue crayon to circle the sticks that are longer than the 5-stick.

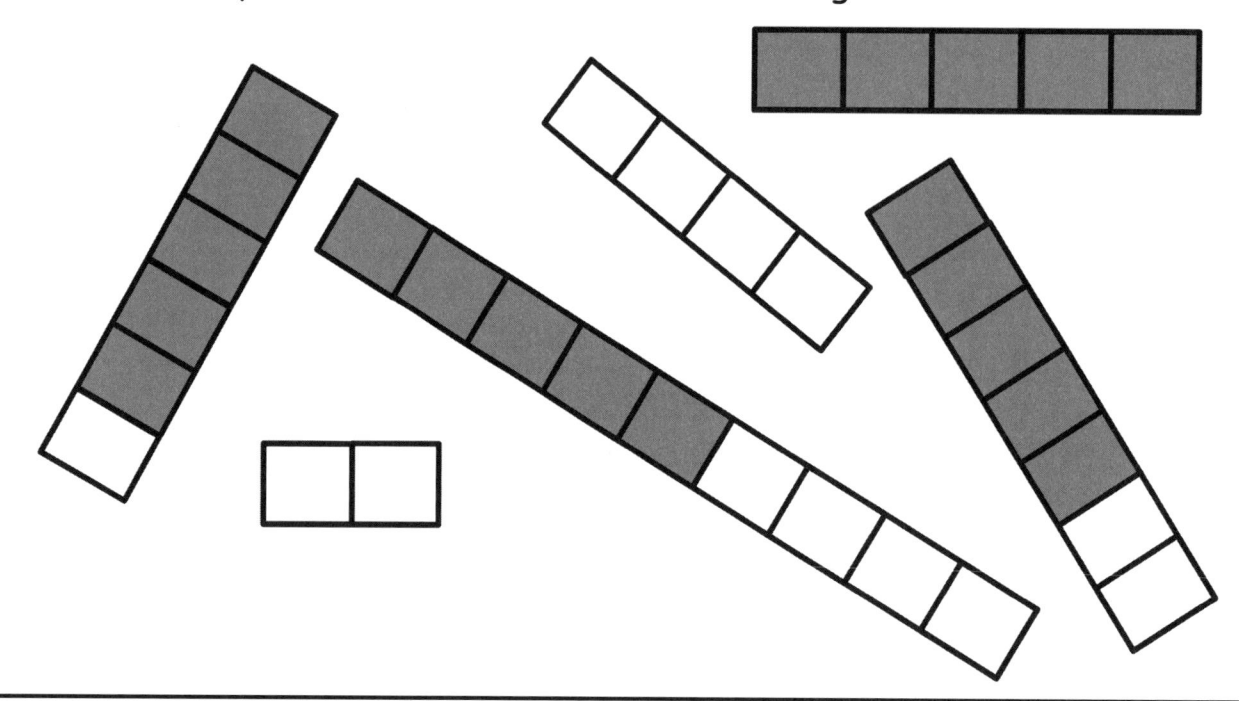

On the back, draw a 7-stick. Draw a stick longer than it. Draw a stick shorter than it.

Longer than my 5-stick:

Shorter than my 5-stick:

longer or shorter mat

Lesson 4: Compare the length of linking cube sticks to a 5-stick.

Lesson 5

Objective: Determine which linking cube stick is *longer than* or *shorter than* the other.

Suggested Lesson Structure

- ■ Fluency Practice   (10 minutes)
- ■ Application Problem   (5 minutes)
- ■ Concept Development   (29 minutes)
- ■ Student Debrief   (6 minutes)
- **Total Time**   **(50 minutes)**

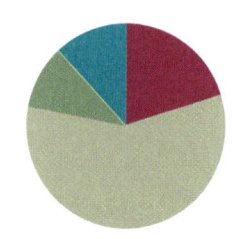

Fluency Practice (10 minutes)

- Show Me Longer and Shorter **K.MD.1**   (2 minutes)
- 5-Group Hands **K.CC.2**   (4 minutes)
- 5-Groups on the Dot Path **K.CC.2**   (4 minutes)

Show Me Longer and Shorter (2 minutes)

Note: This kinesthetic activity reviews vocabulary.

Conduct the activity as described in Lesson 2, but with *longer* and *shorter*. Now, students extend their hands from side to side to indicate length.

5-Group Hands (4 minutes)

Materials: (T) Large 5-group cards in vertical orientation (Fluency Template 1)

Note: This maintenance activity develops flexibility in seeing the 5-groups vertically or horizontally and adds a kinesthetic component.

- T: (Show the 6-dot card in vertical orientation.) Raise your hand when you know how many dots are on the left. (Wait until all hands are raised, and then signal.) Ready?
- S: 5.
- T: Right?
- S: 1.
- T: We can show this 5-group on our hands. Five on the left and 1 on the right, like this. (Demonstrate on hands, side by side.)
- S: (Show 5 and 1 on hands, side by side.)

A STORY OF UNITS   Lesson 5   K•3

T: Push your hands out as you count on from 5, like this. 5 (extend the left hand forward), 6 (extend the right hand forward). Try it with me.

S: 5 (extend the left hand forward), 6 (extend the right hand forward).

Continue with 5, 6, and 7, steadily decreasing guidance from the teacher, until students can show the 5-groups on their hands with ease.

5-Groups on the Dot Path (4 minutes)

Materials: (S) Dot path (Fluency Template 2) placed inside of a personal white board

Note: This activity helps students gain flexibility in grouping 5 and starting to count on from 5 pictorially.

T: Touch and count the dots on your dot path.
S: 1, 2, 3, ..., 10.
T: What do you notice about the dot path?
S: There are 10 dots. → There are two different color dots. → A color change at 5.
T: Yes. I'm going to ask you to circle a group of dots. Use the color change at 5 to count and circle them as fast as you can. Ready? Circle 5.
S: (Circle a group of 5 dots.)
T: How did you do that so fast?
S: I just circled all the light ones, and I knew it was 5.
T: Erase. Get ready for your next number. Circle 6.
S: (Circle a group of 6 dots.)
T: How did you count 6?
S: I counted all of the dots until I got to 6. → I counted one more than 5.

If students are starting to count on, let them share their thinking with the class. Continue the process with numbers to 10. Deviate from a predictable pattern as students show mastery.

Application Problem (5 minutes)

Write your name so that one letter is in each box. Begin with the box above the star. Don't skip any boxes!

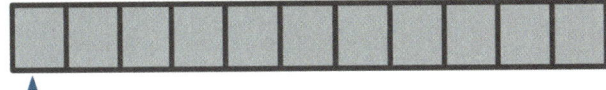

★

You made a name train. Compare your train to that of your partner. What do you notice? Which train has more letter passengers?

Note: By replacing the vertical emphasis in yesterday's linking cube exercise with a horizontal representation, the problem serves as an anticipatory set for today's lesson. Circulate during the discussion to notice use of *longer than* and *shorter than* terminology; observe endpoint alignment skills.

> **NOTES ON MULTIPLE MEANS OF ENGAGEMENT:**
>
> Scaffold the Application Problem for English language learners and students with disabilities by walking them through the directions one step at a time. Begin with the box above the star, and point to it. Observe students as they follow directions to ensure their complete understanding.

Lesson 5: Determine which linking cube stick is *longer than* or *shorter than* the other.

Concept Development (29 minutes)

Materials: (S) 1 bag of linking cube stairs from Lesson 4 per pair

T: With your partner, arrange your linking cube stairs from yesterday on your desk. This time, put them in order from the tallest to shortest. Let's count to make sure they are all here. How many are in the longest stick?
S: 10. 1 less is 9. 9. 1 less is 8. (Continue the pattern through 1.) 1.
T: Find your 5-stick, and hold it up. How many?
S: 5.
T: Now, find your 2-stick, and compare it to your 5-stick. What do you notice?
S: It is shorter.
T: Repeat after me, "My 2 is shorter than my 5. My 5 is longer than my 2." (Hold up sticks, and demonstrate.)
S: My 2 is shorter than my 5. My 5 is longer than my 2.
T: Put your sticks down. Find your 5 and your 4. Compare the sticks. What do you notice?
S: My 4 is shorter than my 5. → My 5 is longer than my 4.
T: Great! Now, use your 5, and choose another stick of your own. What did you choose?
S: The 7. It is longer than the 5. (Answers may vary.)
T: Say it with me. "The 7 is longer than the 5. The 5 is shorter than the 7." Did anyone choose a different stick? (Allow other students to tell about their choices.)

Repeat this exercise and sentence modeling through several iterations, using a variety of different sticks for the initial comparison.

T: Do you see a stick that is shorter than the 1? Why not? (Allow time for discussion.)
T: Do you see a stick that is longer than the 10? Why not? (Allow time for discussion.)

MP.2

T: Mix up all of your sticks on your desk. Now, you will play a game with your partner. One of you will close your eyes and choose two sticks. When you open your eyes, quickly tell your partner which stick is longer than the other one and which stick is shorter than the other one. Make sure you tell your partner in the way that we just practiced! Then, it will be your partner's turn. (Allow students to play until they are comfortable with the correct language of comparison.)

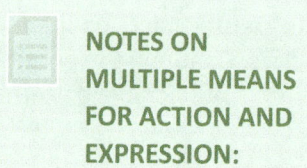

> **NOTES ON MULTIPLE MEANS FOR ACTION AND EXPRESSION:**
>
> When giving directions about how to play the game, illustrate the meaning for English language learners. Hold up two sticks and demonstrate by saying, "My 2-stick is shorter than my 7-stick. My 7-stick is longer than my 2-stick."

T: What did you notice while you were playing your game? (Allow time for responses.)
T: Did it matter if your sticks were up, down, or sideways?
S: No! They were still the same length!
T: Put your stairs away carefully for next time.

Problem Set (10 minutes)

Students should do their personal best to complete the Problem Set within the allotted time.

For this Problem Set, it is recommended that all students begin with circling the sticks and possibly leave filling in the blanks to the end if time allows.

Student Debrief (6 minutes)

Lesson Objective: Determine which linking cube stick is *longer than* or *shorter than* the other.

The Student Debrief is intended to invite reflection and active processing of the total lesson experience.

Invite students to review their solutions for the Problem Set. They should check work by comparing answers with a partner before going over answers as a class. Look for misconceptions or misunderstandings that can be addressed in the Debrief. Guide students in a conversation to debrief the Problem Set and process the lesson.

Any combination of the questions below may be used to lead the discussion.

- When you were asked to draw a stick taller or shorter than 6 or 9 cubes, what did you draw?
- Did you all choose to draw the same stick? What else could you have chosen?
- How did you compare the lengths of your sticks?
- Tell your partner about the number of cubes in the stick you drew using the sentence, "My _____ stick is longer than/shorter than my _____ stick."
- Turn to your partner, and tell him something you could teach or share with your family tonight about length. Be sure to use the words *longer than* and *shorter than*.

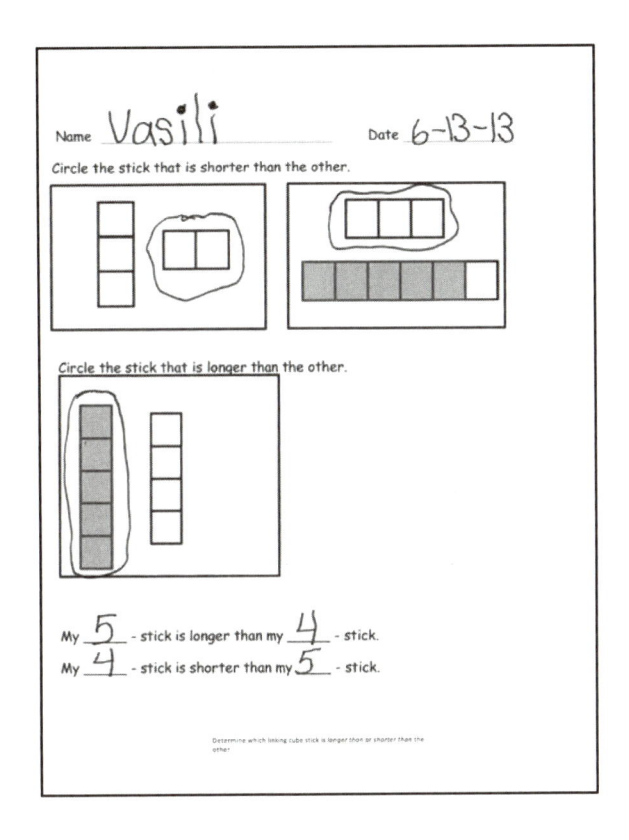

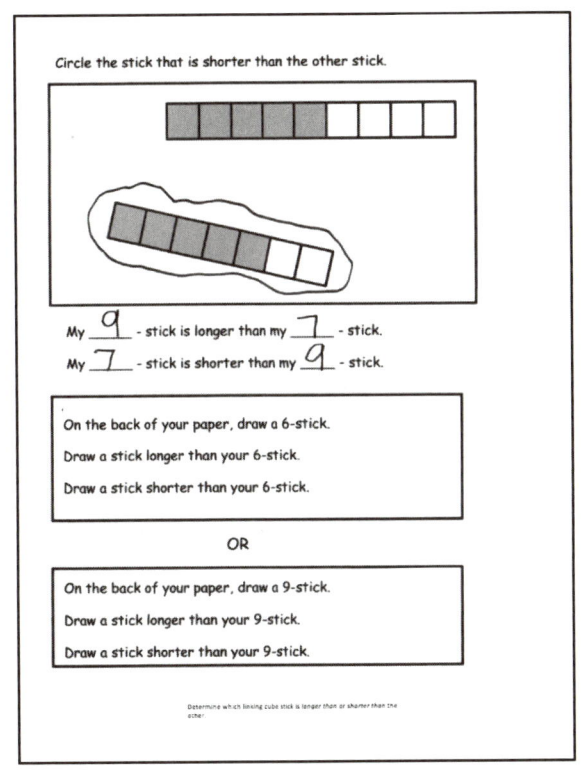

A STORY OF UNITS   Lesson 5 Problem Set K•3

Name _____   Date _____

Circle the stick that is shorter than the other.

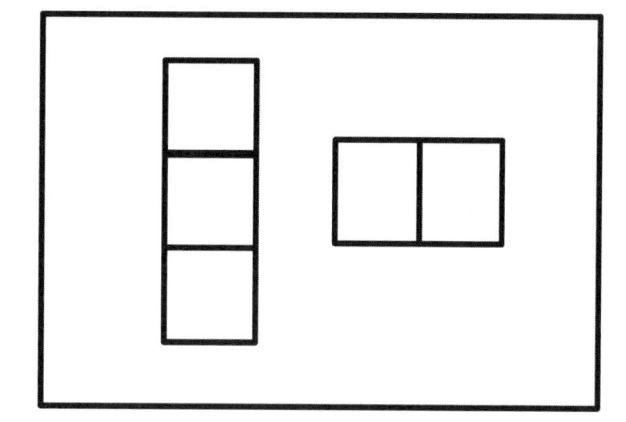

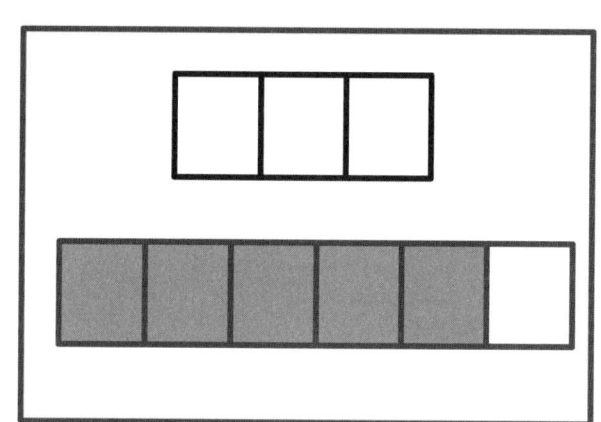

Circle the stick that is longer than the other.

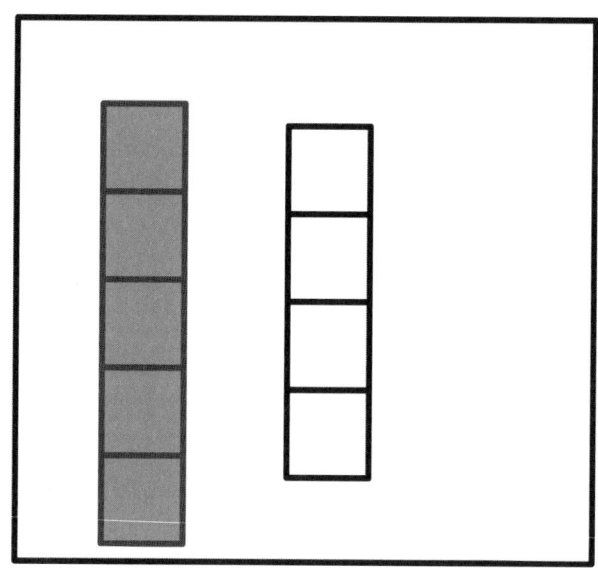

My _____ -stick is longer than my _____ -stick.

My _____ -stick is shorter than my _____ -stick.

Lesson 5:   Determine which linking cube stick is *longer than* or *shorter than* the other.

A STORY OF UNITS   Lesson 5 Problem Set K•3

Circle the stick that is shorter than the other stick.

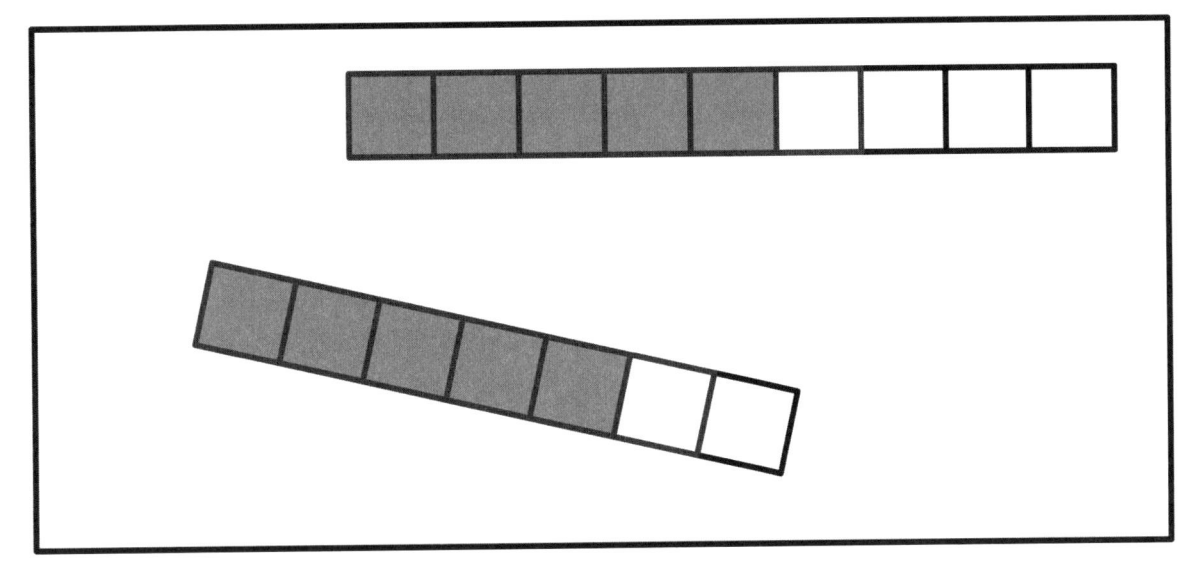

My _____ -stick is longer than my _____ -stick.

My _____ -stick is shorter than my _____ -stick.

On the back of your paper, draw a 6-stick.

Draw a stick longer than your 6-stick.

Draw a stick shorter than your 6-stick.

OR

On the back of your paper, draw a 9-stick.

Draw a stick longer than your 9-stick.

Draw a stick shorter than your 9-stick.

Lesson 5:   Determine which linking cube stick is *longer than* or *shorter than* the other.

A STORY OF UNITS

Lesson 5 Homework K•3

Name _____ Date _____

Circle the stick that is shorter than the other.

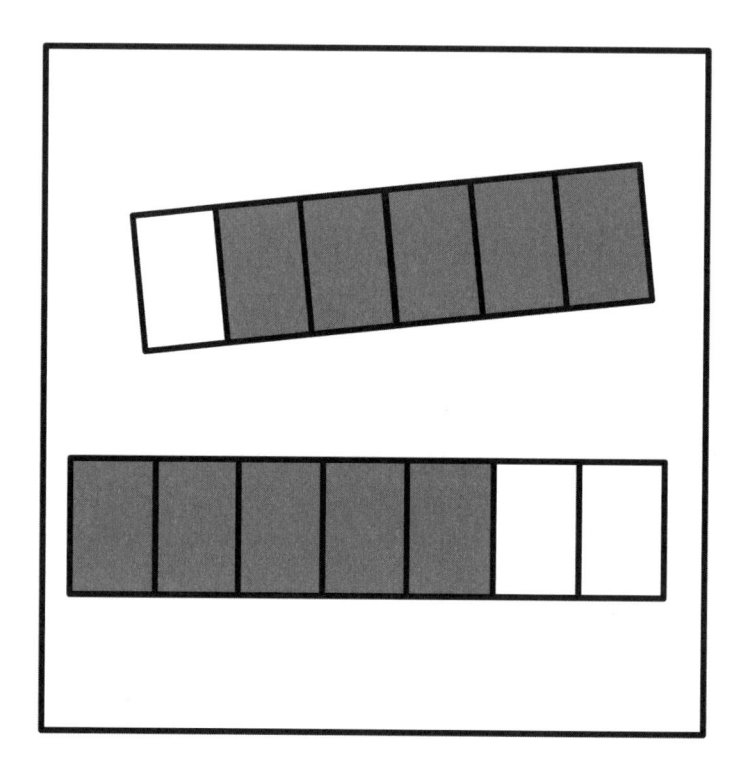

My _____ -stick is shorter than my _____ -stick.

My _____ -stick is longer than my _____ -stick.

On the back of your paper, draw a 7-stick.

Draw a stick that is longer than the 7-stick.

Draw a stick that is shorter than the 7-stick.

Lesson 5: Determine which linking cube stick is *longer than* or *shorter than* the other.

A STORY OF UNITS

Lesson 5 Homework K•3

Circle the stick that is longer than the other.

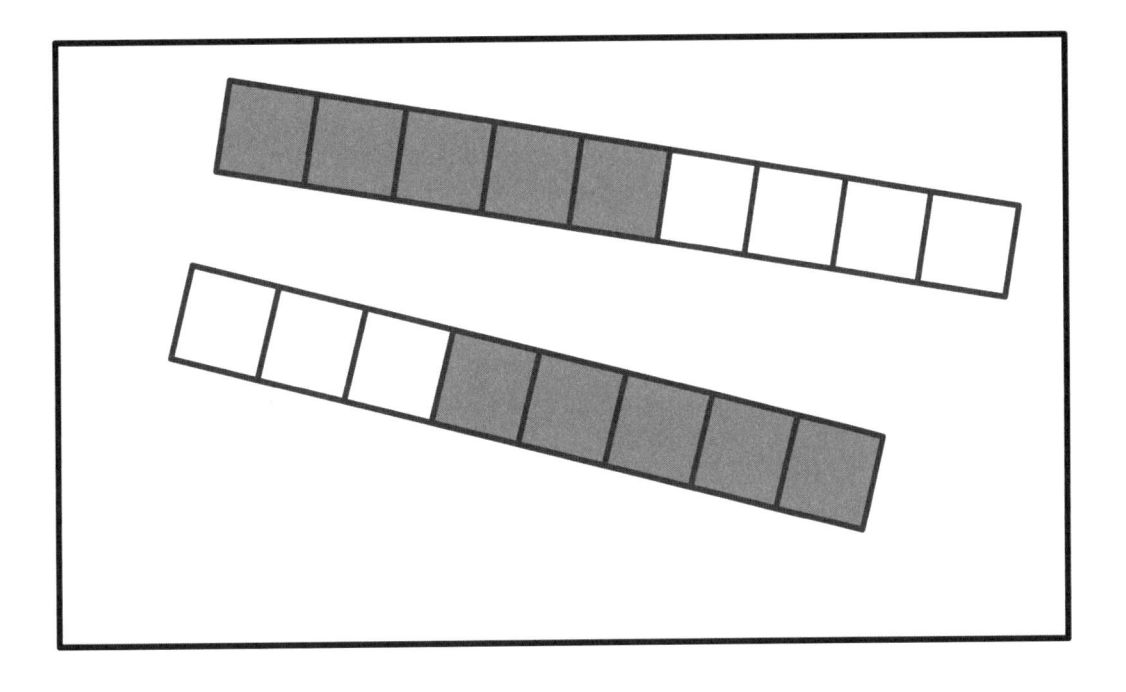

My _____ -stick is shorter than my _____ -stick.

My _____ -stick is longer than my _____ -stick.

On the back of your paper, draw a stick that is between a 4- and a 6-stick.

Draw a stick that is longer than your new stick.

Draw a stick this is shorter than your new stick.

Lesson 5: Determine which linking cube stick is *longer than* or *shorter than* the other.

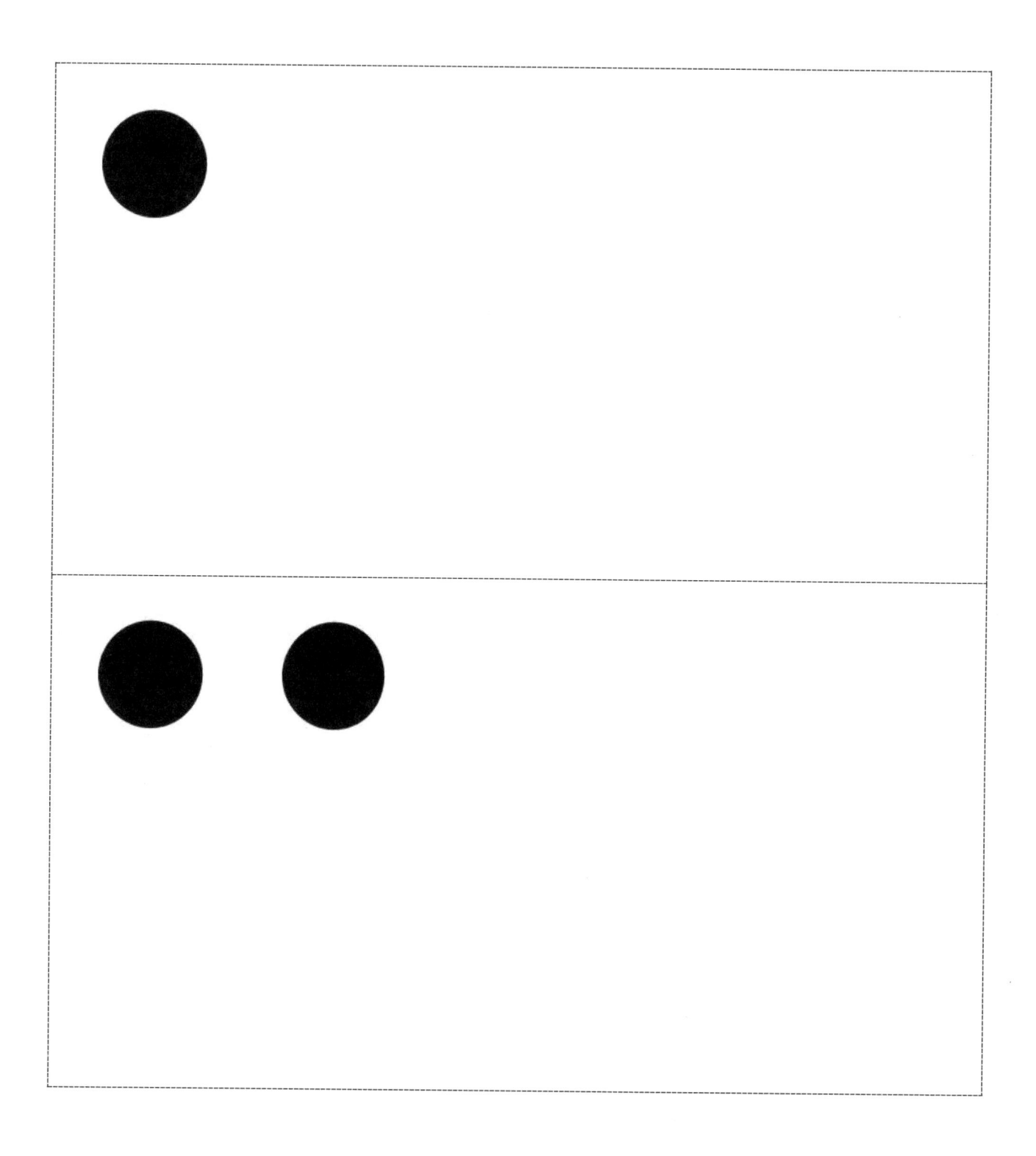

large 5-group cards (Copy on card stock and cut. Save a full set.)

Lesson 5: Determine which linking cube stick is *longer than* or *shorter than* the other.

A STORY OF UNITS

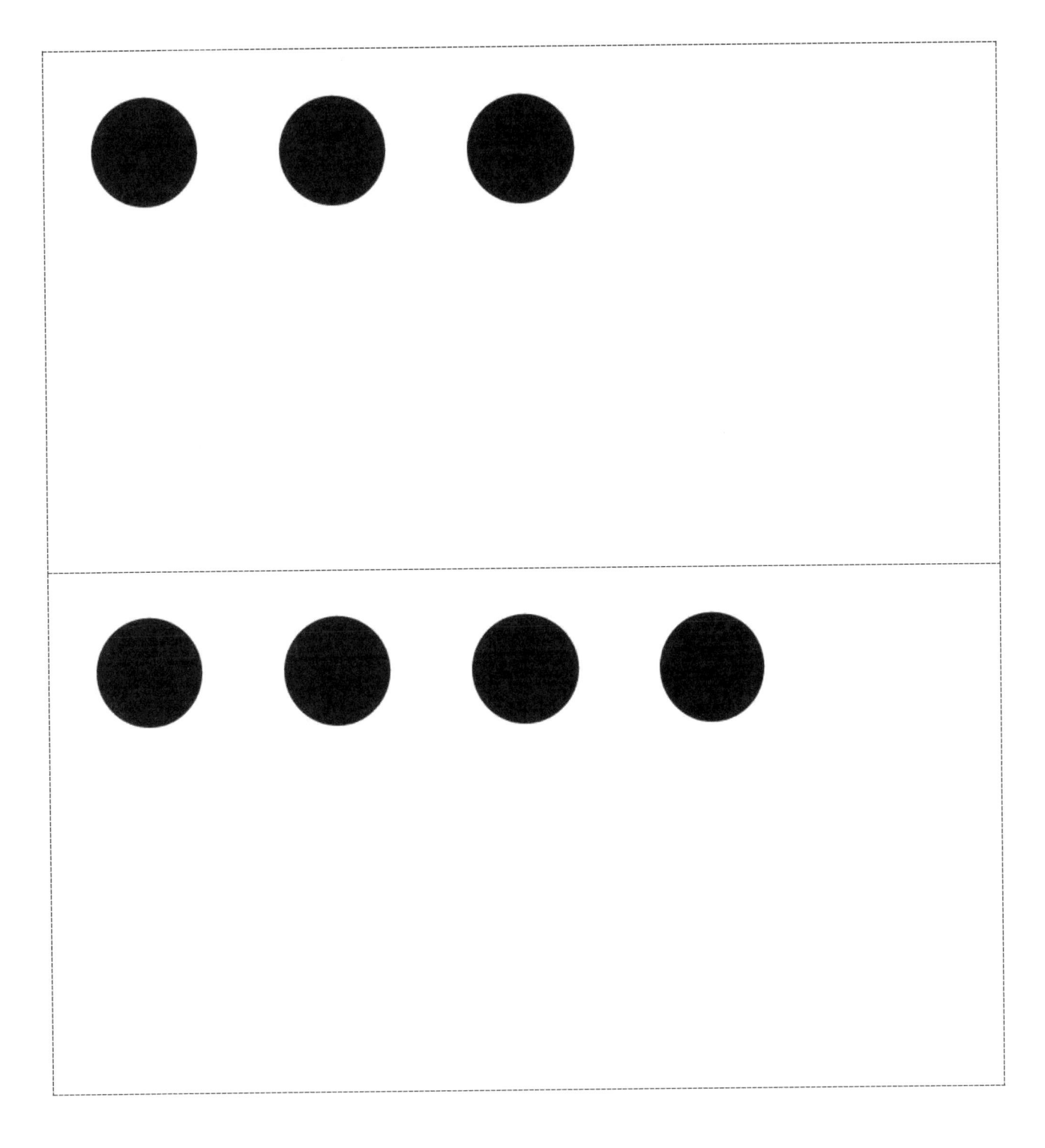

Lesson 5 Fluency Template 1   K•3

large 5-group cards (Copy on card stock and cut. Save a full set.)

Lesson 5: Determine which linking cube stick is *longer than* or *shorter than* the other.

57

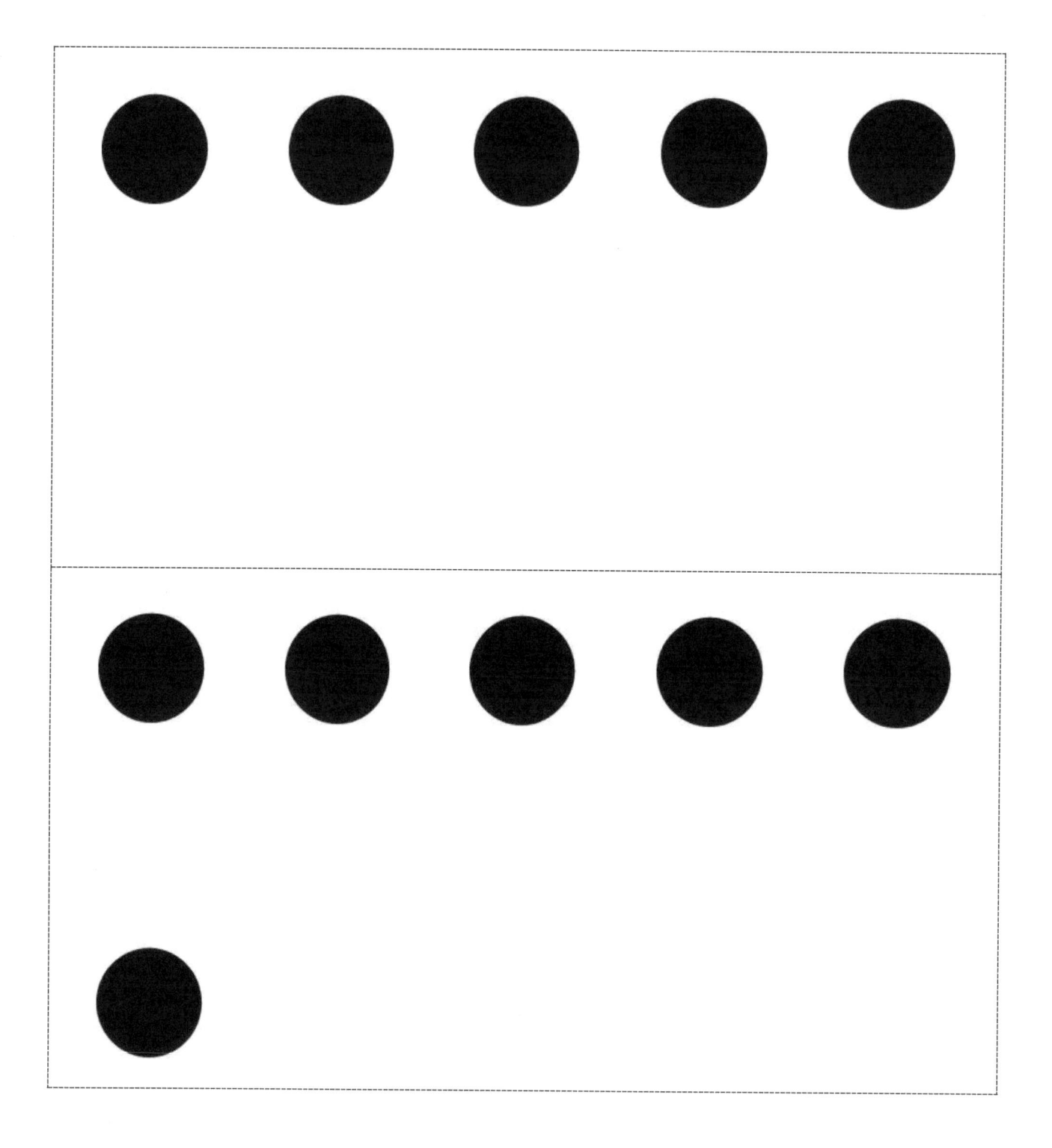

large 5-group cards (Copy on card stock and cut. Save a full set.)

Lesson 5: Determine which linking cube stick is *longer than* or *shorter than* the other.

A STORY OF UNITS

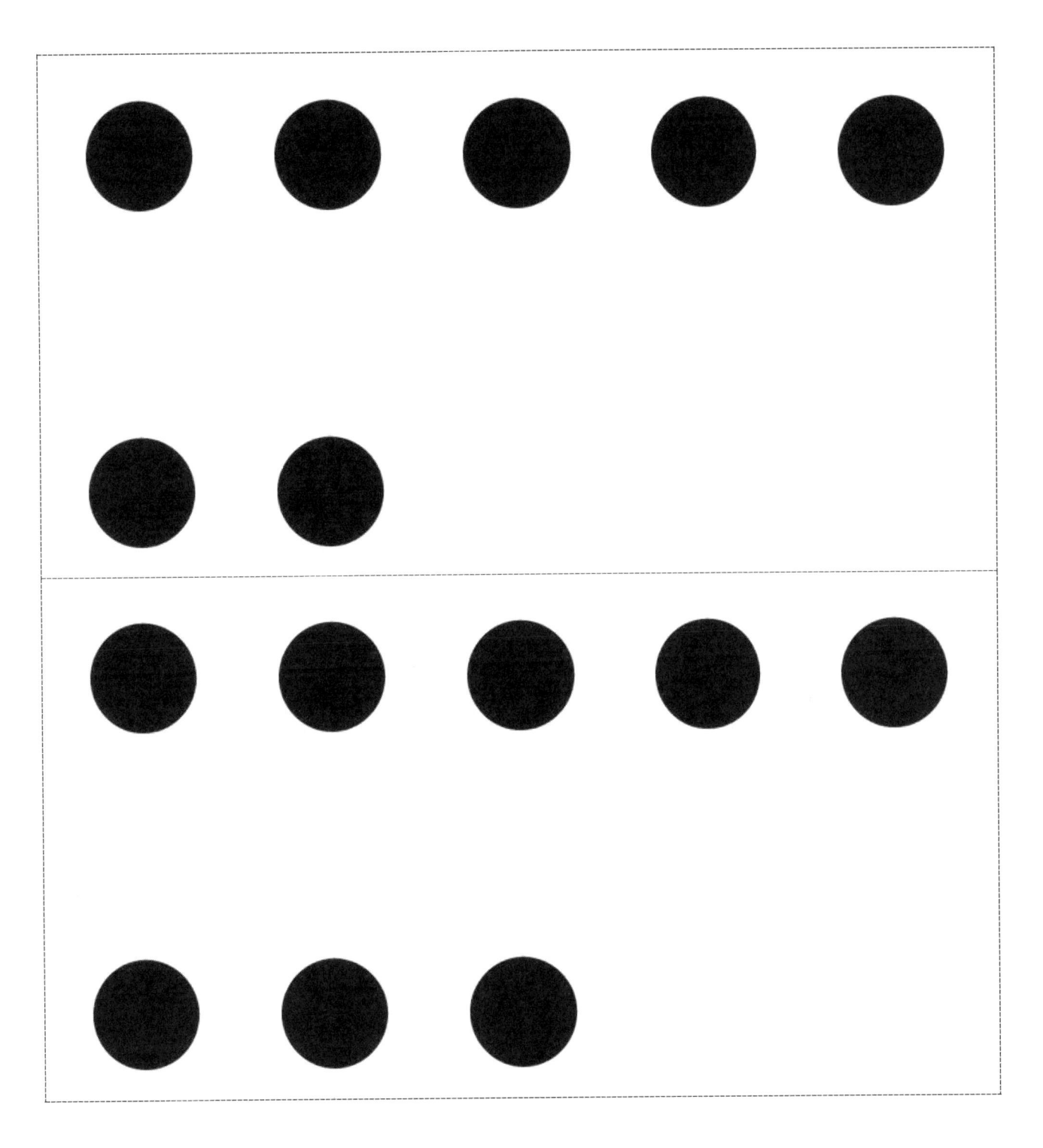

Lesson 5 Fluency Template 1   K•3

large 5-group cards (Copy on card stock and cut. Save a full set.)

EUREKA MATH

Lesson 5: Determine which linking cube stick is *longer than* or *shorter than* the other.

© 2015 Great Minds. eureka-math.org
GK-M3-TE-B3-1.3.1-01.2016

59

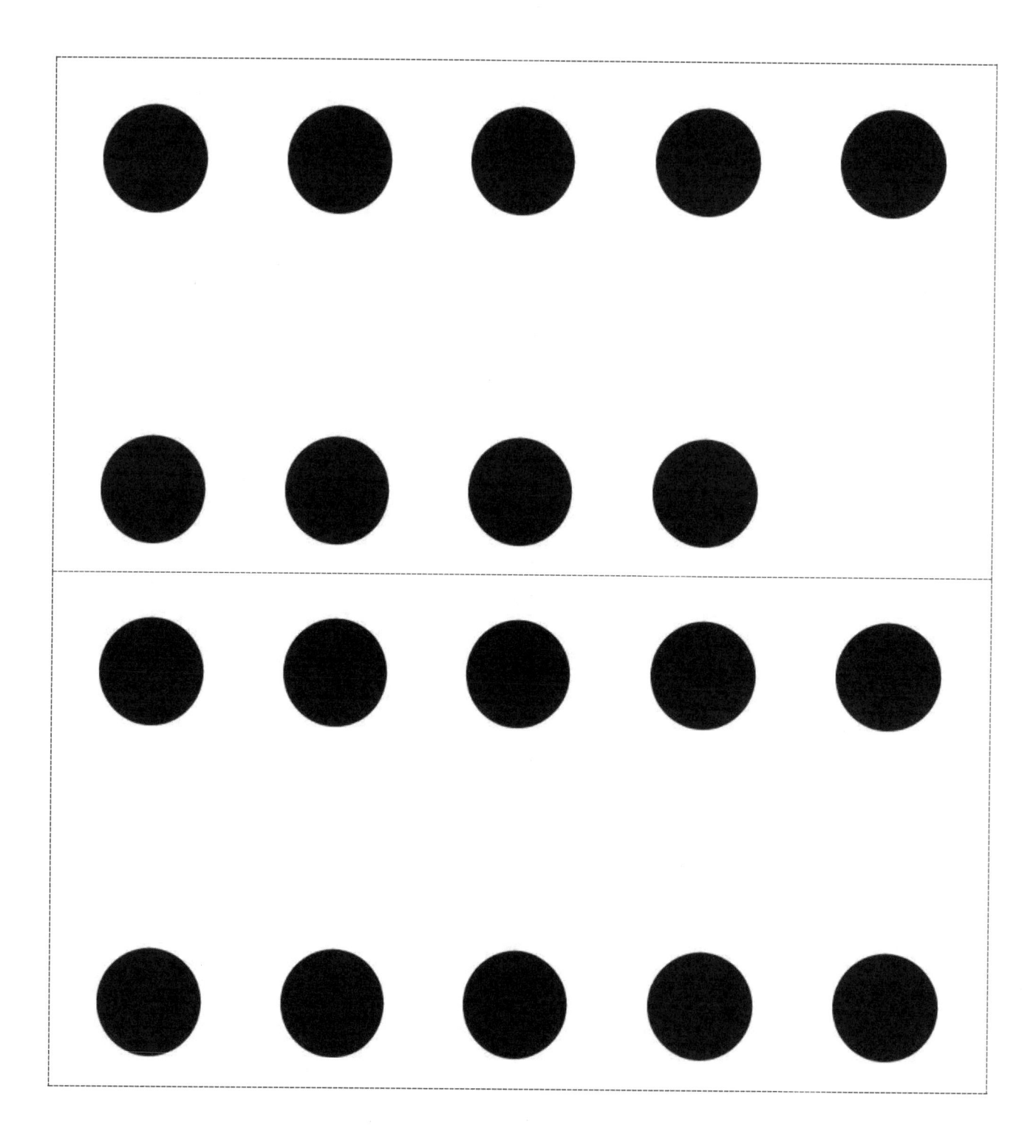

large 5-group cards (Copy on card stock and cut. Save a full set.)

Lesson 5: Determine which linking cube stick is *longer than* or *shorter than* the other.

A STORY OF UNITS

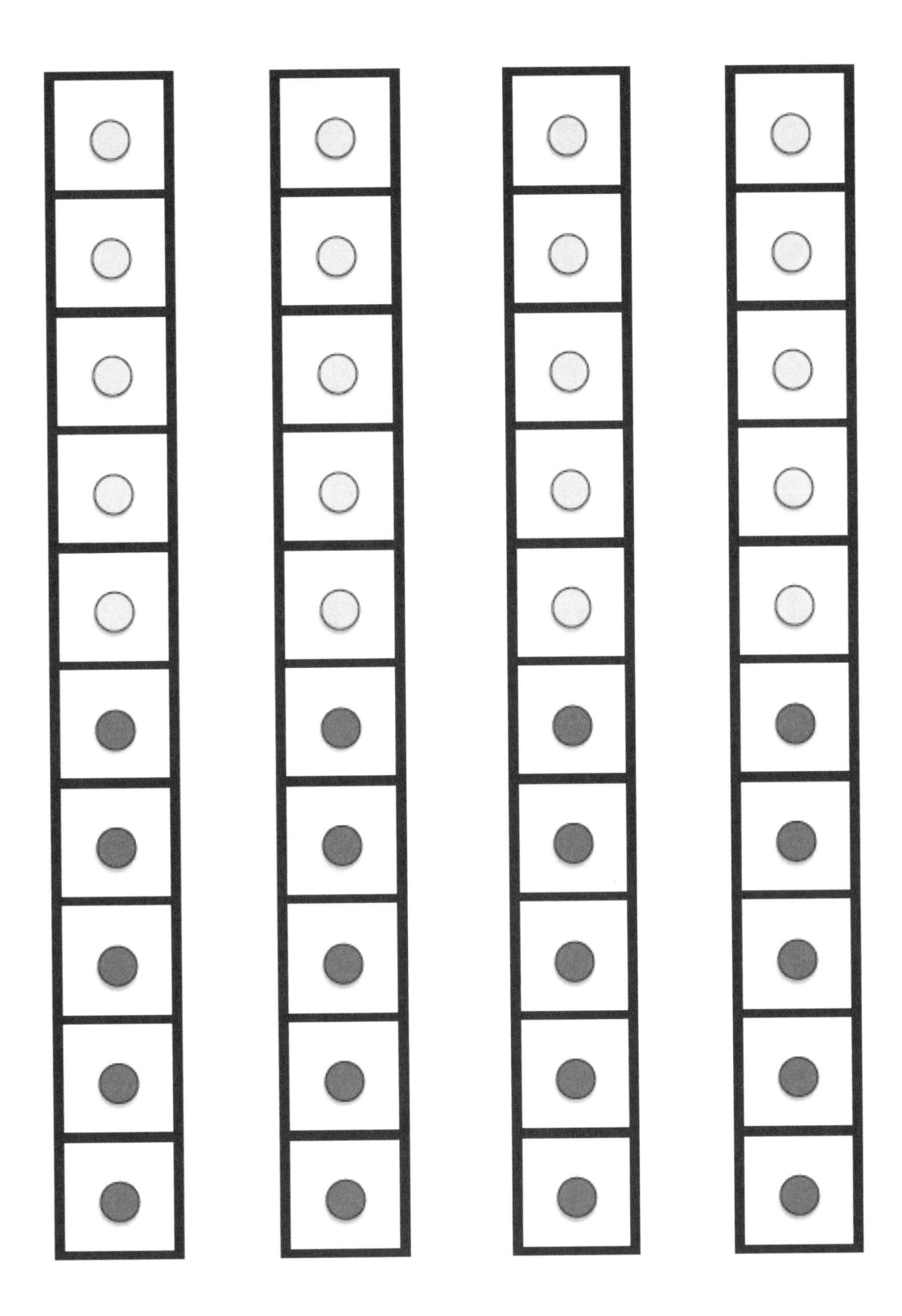

Lesson 5 Fluency Template 2   K•3

dot path

Lesson 5: Determine which linking cube stick is *longer than* or *shorter than* the other.

EUREKA MATH

© 2015 Great Minds. eureka-math.org
GK-M3-TE-B3-1.3.1-01.2016

61

Lesson 6

Objective: Compare the length of linking cube sticks to various objects.

Suggested Lesson Structure

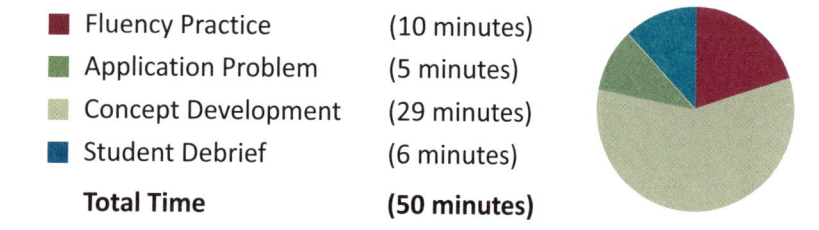

- ■ Fluency Practice   (10 minutes)
- ■ Application Problem   (5 minutes)
- ■ Concept Development   (29 minutes)
- ■ Student Debrief   (6 minutes)
- **Total Time**   **(50 minutes)**

Fluency Practice (10 minutes)

- Show Me Taller and Shorter **K.MD.1**   (3 minutes)
- Counting the Say Ten Way with the Rekenrek **K.NBT.1**   (4 minutes)
- Hidden Numbers **K.OA.3**   (3 minutes)

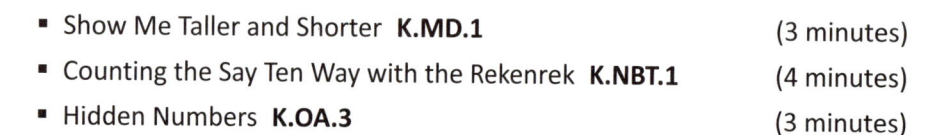

Show Me Taller and Shorter (3 minutes)

Note: This kinesthetic fluency activity reviews vocabulary.

Conduct the activity as described in Lesson 2, but with *longer* and *shorter*. Now, students extend their hands from side to side to indicate length.

Counting the Say Ten Way with the Rekenrek (4 minutes)

Materials: (T) 20-bead Rekenrek

Note: This activity is an extension of students' previous work with the Rekenrek and anticipates working with teen numbers.

- T: We can count with the Rekenrek the same way we do our Say Ten *push-ups* (fluency activity in Lesson 3). (Keep the screen on the right side of the Rekenrek to cover the beads that are not being counted. Slide over all of the beads on the top row.) How many do you see?
- S: 10.
- T: Here's 1 more. (Slide over 1 bead on the bottom row.) That's what ten 1 looks like on the Rekenrek. How many do you see?
- S: Ten 1.
- T: (Slide 1 more bead over on the bottom row.) How many do you see?
- S: Ten 2.
- T: (Slide 1 more bead over on the bottom row.) How many do you see?
- S: Ten 3.

Continue counting forward and backward with the following suggested sequence: ten 1, ten 2, ten 1, ten 2, ten 3, ten 2, ten 3, ten 2, ten 1.

Hidden Numbers (3 minutes)

Materials: (S) Hidden numbers mat (Lesson 3 Fluency Template) inserted into personal white board

Note: Finding embedded numbers anticipates the work of Kindergarten Module 4 by developing part–whole thinking.

Conduct the activity as described in Lesson 3, but this time, guide students to find hidden numbers within a group of 7.

Application Problem (5 minutes)

Materials: (S) Crayon, paper, bag of linking cube stairs

Spread your hand out on the piece of paper, and trace around it to make your handprint. Now, take your hand off of the paper, and look carefully at the fingers in your handprint drawing.

Think about which linking cube stick might be as long as your thumb. Take out that stick, and check your guess. Were you right? Which one would be about as long as your little finger? Your middle finger?

NOTES ON MULTIPLE MEANS OF REPRESENTATION:

Scaffold the Application Problem for struggling students by first modeling and then helping them trace their hands. Watch as they follow the first direction until it is ensured that they are clear about what to do.

Test your guesses to see if you were close. Share your discoveries with your friend. Are your friend's fingers and your fingers the same lengths?

Note: This problem gives students experience using linking cubes as a simple comparison tool in anticipation of today's lesson.

Concept Development (29 minutes)

Materials: (S) Bag of linking cube number stairs and paper bag filled with various items to measure (e.g., pencil, eraser, glue stick, toy car, small block, 8-inch piece of string, marker, child's scissors, crayon) per pair

Note: Save a set of student materials for the culminating task in Lesson 32.

MP.3

T: With your partner, take the items out of your mystery bag, and place them on your desk. Now, use the linking cube sticks to make a set of number stairs on your desk. Put them in order from the 1-cube to the 10-stick. Let's count to make sure we have them all.

S: 1. 1 more is 2. 2. 1 more is 3. (Continue the pattern through 10.) 10.

T: Find the crayon. Hold it up. Now, guess which cube stick might be the same length as your crayon. Test your guess with your partner. (Allow time for discussion and comparison of the lengths.)

T: Find your 10-stick. Look at the items from your mystery bag. Point to something that you think might be shorter than your 10-stick. Now, compare the length of your 10-stick with the length of your object. Test your guess. Were you correct? (Allow time for discussion and comparison of the lengths.)

S: Yes!

Lesson 6: Compare the length of linking cube sticks to various objects.

A STORY OF UNITS Lesson 6 K•3

MP.3

T: This time, point to something that you think will be longer than your 4-stick. Test your guess. Were you correct?

S: Yes!

T: We will play Simon Says. Simon says, point to something that you think is shorter than your 8-stick. Simon says, test your guess. Simon says, hold up your object if you were correct. Put it down. I didn't say Simon says!

T: Simon says, point to something that you think is taller than your 2-stick. Simon says, test your guess. Hold up your object if you were correct. I didn't say Simon says!

T: (Continue playing the game several times, varying the *shorter than, longer than,* and *taller than* language and incorporating all of the number sticks at least once. Observe accuracy of student responses with respect to object length.)

T: Great listening! Put your objects back in your mystery bag, and carefully put away the linking cube sticks. We will be talking more about linking cube sticks during our Problem Set.

> **NOTES ON MULTIPLE MEANS FOR ACTION AND EXPRESSION:**
>
> English language learners are better able to follow the fast pace of the Simon Says game if practiced a few times with familiar commands. Simon says, raise your hand. Simon says, sit down.

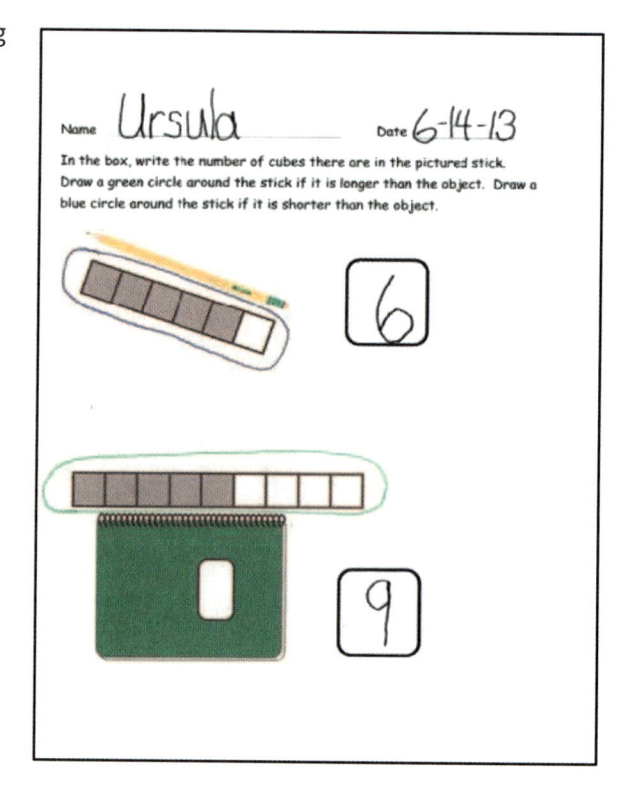

Problem Set (10 minutes)

Note: The problems in today's Problem Set may take more time than usual since reading and writing are required. The problems are meant to be done with the teacher reading the sheet and guiding the students through the Problem Set. Students should do their personal best to complete the Problem Set within the allotted time.

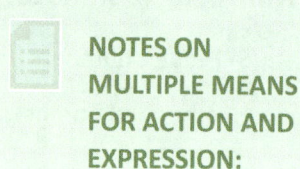

64 Lesson 6: Compare the length of linking cube sticks to various objects.

Student Debrief (6 minutes)

Lesson Objective: Compare the length of linking cube sticks to various objects.

The Student Debrief is intended to invite reflection and active processing of the total lesson experience.

Invite students to review their solutions for the Problem Set. They should check work by comparing answers with a partner before going over answers as a class. Look for misconceptions or misunderstandings that can be addressed in the Debrief. Guide students in a conversation to debrief the Problem Set and process the lesson.

Any combination of the questions below may be used to lead the discussion.

- What did we do when we were playing Simon Says?
- How did you make your guesses?
- What did you draw on your Problem Set that was longer than your 3-stick? Shorter than your 3-stick?
- Can you think of something at home that would be shorter than your 5-stick? Bring it tomorrow so that you may test your guess!

A STORY OF UNITS

Lesson 6 Problem Set K•3

Name _____ Date _____

In the box, write the number of cubes there are in the pictured stick.
Draw a green circle around the stick if it is longer than the object.
Draw a blue circle around the stick if it is shorter than the object.

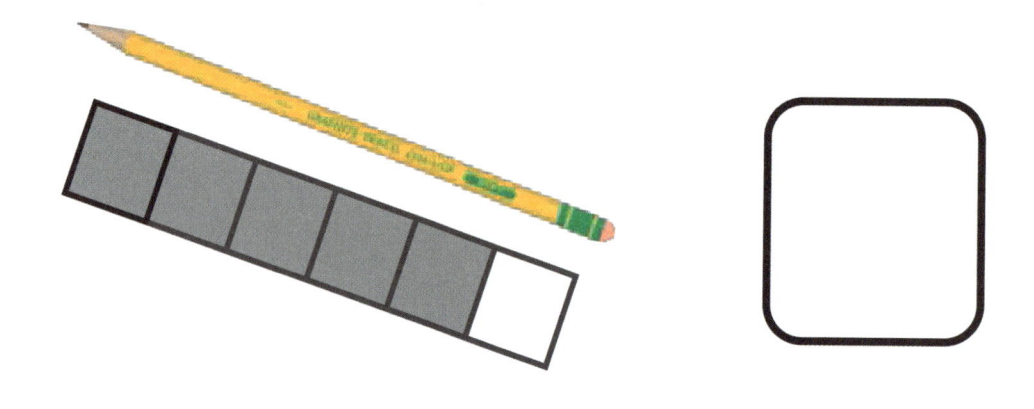

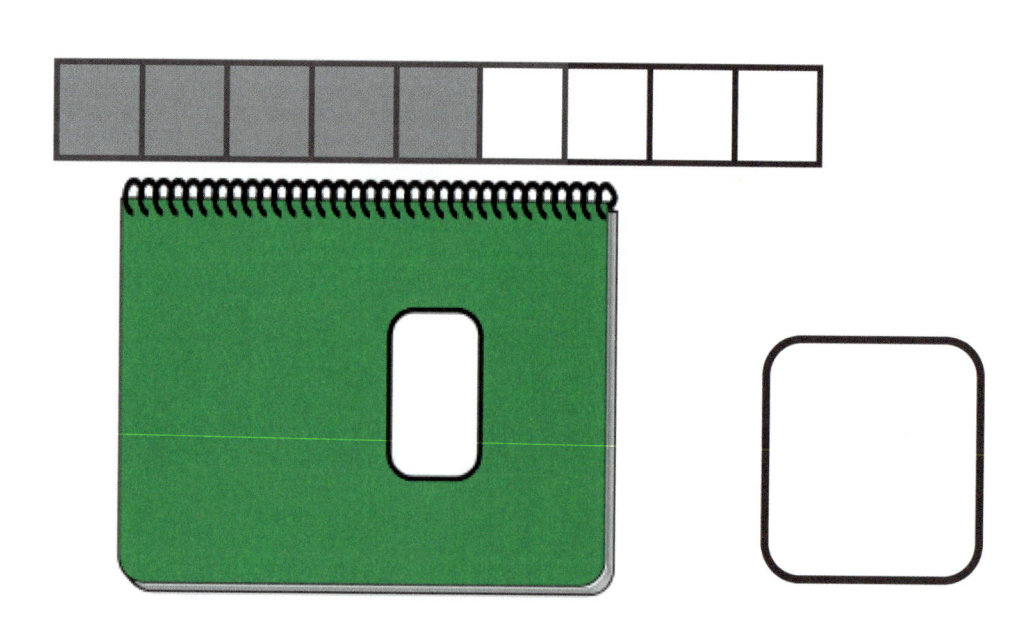

Lesson 6: Compare the length of linking cube sticks to various objects.

Lesson 6 Problem Set

Make a 3-stick. In your classroom, select a crayon, and see if your crayon is longer than or shorter than your stick.

Trace your 3-stick and your crayon to compare their lengths.

In your classroom, find a marker, and make a stick that is longer than your marker.

Trace your stick and your marker to compare their lengths.

Make a 5-stick. Find something in the classroom that is longer than your 5-stick.

Trace your 5-stick and the object to compare their lengths.

Lesson 6: Compare the length of linking cube sticks to various objects.

A STORY OF UNITS

Lesson 6 Homework K•3

Name _____ Date _____

Color the cubes to show the length of the object.

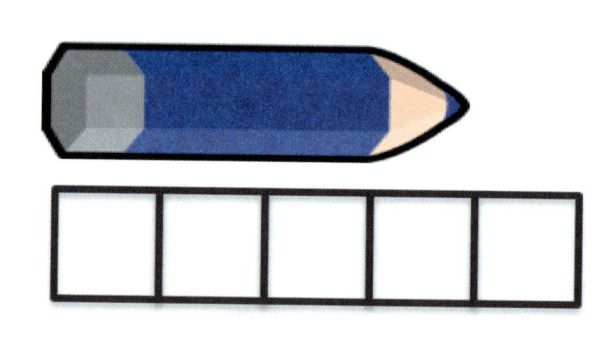

Lesson 6: Compare the length of linking cube sticks to various objects.

Lesson 7

Objective: Compare objects using *the same as*.

Suggested Lesson Structure

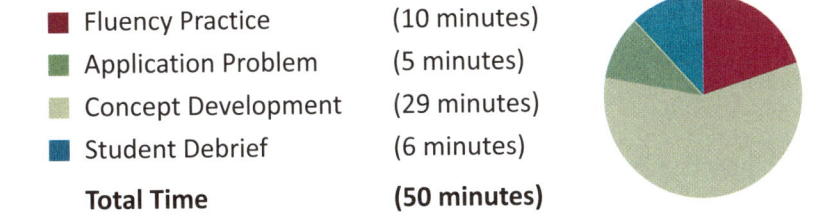

- **Fluency Practice** (10 minutes)
- **Application Problem** (5 minutes)
- **Concept Development** (29 minutes)
- **Student Debrief** (6 minutes)
- **Total Time** **(50 minutes)**

Fluency Practice (10 minutes)

- Counting the Say Ten Way with the Rekenrek **K.NBT.1**   (3 minutes)
- Roll and Draw 5-Groups **K.OA.3**   (4 minutes)
- Green Light, Red Light **K.CC.2**   (3 minutes)

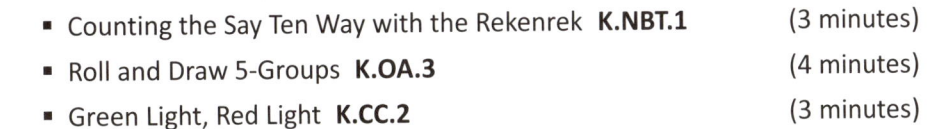

Counting the Say Ten Way with the Rekenrek (3 minutes)

Conduct activity as described in Lesson 6, but this time, continue to ten 5.

Note: This activity is an extension of students' previous work with the Rekenrek and anticipates working with teen numbers.

Roll and Draw 5-Groups (4 minutes)

Materials: (S) Die (with the 6-dot side covered), personal white board

Note: Observe to see which students erase completely and begin each time from 1 rather than draw more or erase some to adjust to the new number. By drawing 5-groups, students see numbers in relationship to the five.

Roll the die, count the dots, and then draw the number as a 5-group.

Green Light, Red Light (3 minutes)

Materials: (T) Green and red dry-erase markers

On the board, draw a green dot with a 1 underneath and a red dot with a 3 underneath. Explain to students that they should start counting and stop counting on the number as indicated by the color code.

- T: Look at the numbers. (Point to the number 1 written below the green dot and the number 3 below the red dot.) Think. Ready? Green light!
- S: 1, 2, 3.

T: Very good! (Erase numbers 1 and 3, and write the new numbers.) New numbers (green is 3, red is 1). Look, think, ready... green light!

S: 3, 2, 1.

At this point in the year, it may not be necessary to start at 1. Work within a range that is comfortable for the students, and build incrementally. Challenge them by frequently changing directions between counting up and counting down.

Application Problem (5 minutes)

Materials: (S) Small ball of clay

Make a little clay snake that is as long as your pointer finger. Now, make a friend for him that is as long as your pinky finger. Which one is longer? Show your creations to your partner.

Note: Circulate to notice students' strategies in terms of comparing the length of their snakes to their fingers. Tactile creation of objects of equal length to fingers provides the anticipatory set for the more abstract length equality exercise in today's Concept Development.

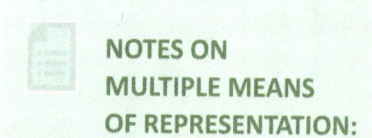

NOTES ON MULTIPLE MEANS OF REPRESENTATION:

Challenge students working above grade level to write down what strategies they used to compare the length of their snakes to their fingers and how they know which snake is longer or shorter than their fingers. Encourage them not only to draw pictures, but also to write the math words they know.

Concept Development (29 minutes)

Materials: (T/S) Bag of linking cube number stairs, riddle work mat (Template) copied on two sides of the paper or inserted into personal white board

MP.7

T: Mix up your number stairs on your desk. Find your 5-stick. Look at it carefully. Now, listen to my riddle. We are two different sticks. We are each shorter than the 5-stick, but when you put us together, we are **the same** length **as** the 5-stick!

T: Which sticks could the riddle be talking about? Look at your sticks, and find two that would work. (Allow time for experimenting.)

T: Student A, what did you find?

S: My 3-stick and my 2-stick. (Hold up sticks.)

T: Right! We would say it like this, "Together, my 3-stick and my 2-stick are the same length as my 5-stick." Repeat after me.

S: Together, my 3-stick and my 2-stick are the same length as my 5-stick.

T: Did anyone do it differently?

S: I found 1 and 4. (Hold up sticks.)

T: Say with me, "Together, my 1-stick and my 4-stick are the same length as my 5-stick."

T: Let's record what we just found. On your work mat, trace your 5-stick like this. (Demonstrate.) Now, trace the 1-stick and the 4-stick underneath the 5-stick you drew. (Demonstrate.) Finish the sentence frame: "Together, my 1-stick and my 4-stick are the same length as my 5-stick."

Lesson 7: Compare objects using *the same as*.

T: We are going to see how many sets of sticks we can find that will make our riddle true. (Allow time for experimenting and recording.)

T: How many different ways did you find to make a stick the same length as your 5-stick? Would anyone like to share their work? (Allow time for discussion and sharing.)

S: I found 4 and 1. → I found 3 and 2. → 2 and 3 works, too!

Problem Set (10 minutes)

Note: Before beginning the Problem Set, take a few minutes to build a stick with students, emphasizing the language in the Problem Set that is continued from the lesson.

Students should do their personal best to complete the Problem Set within the allotted time.

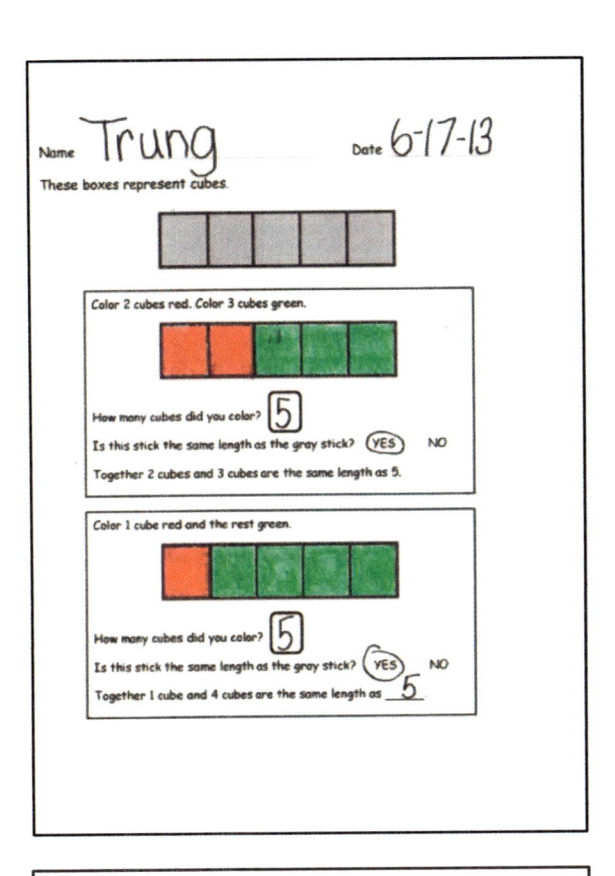

Student Debrief (6 minutes)

Lesson Objective: Compare objects using *the same as.*

The Student Debrief is intended to invite reflection and active processing of the total lesson experience.

Invite students to review their solutions for the Problem Set. They should check work by comparing answers with a partner before going over answers as a class. Look for misconceptions or misunderstandings that can be addressed in the Debrief. Guide students in a conversation to debrief the Problem Set and process the lesson.

Any combination of the questions below may be used to lead the discussion.

- When you made the clay snake today, how could you tell it was **the same** length **as** your finger?
- How did you solve the riddle in the lesson today?
- How did you use the cube sticks to help you solve the riddle?
- Are there other riddles that you can think of about cube sticks that together make the same length as another? Turn to your partner, and see if you can think of some other riddles.

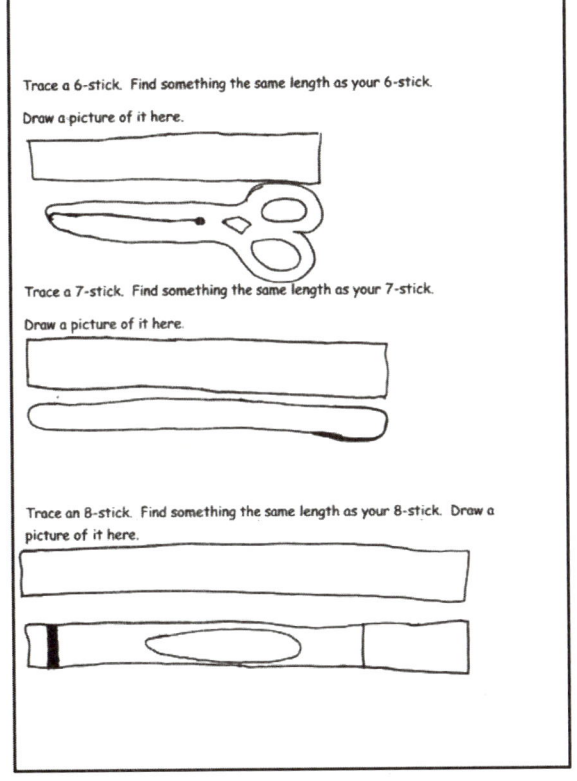

Lesson 7: Compare objects using *the same as.*

A STORY OF UNITS　　　　　　　　　　　　　　　　　　Lesson 7 Problem Set　K•3

Name _____ Date _____

These boxes represent cubes.

Color 2 cubes red. Color 3 cubes green.

How many cubes did you color? ☐

Is this stick the some length as the gray stick?   YES   NO

Together 2 cubes and 3 cubes are the same length as 5.

Color 1 cube red and the rest green.

How many cubes did you color? ☐

Is this stick the same length as the gray stick?   YES   NO

Together 1 cube and 4 cubes are the same length as_____.

Lesson 7:   Compare objects using *the same as*.

Lesson 7 Problem Set

Trace a 6-stick. Find something the same length as your 6-stick. Draw a picture of it here.

Trace a 7-stick. Find something the same length as your 7-stick. Draw a picture of it here.

Trace an 8-stick. Find something the same length as your 8-stick. Draw a picture of it here.

Lesson 7: Compare objects using *the same as*.

A STORY OF UNITS Lesson 7 Homework K•3

Name _____ Date _____

These boxes represent cubes.

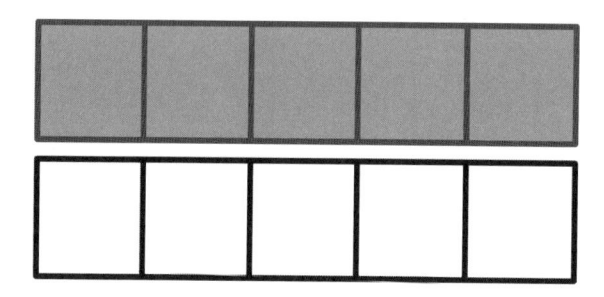

Color 2 cubes green. Color 3 cubes blue.

Together, my green 2-stick and blue 3-stick are the same length as 5 cubes.

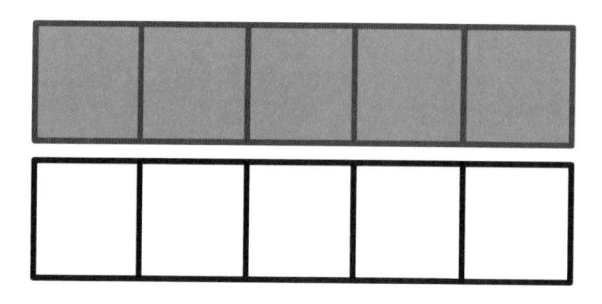

Color 3 cubes blue. Color 2 cubes green.

Together, my blue 3-stick and green 2-stick are the same length as ___ cubes.

Lesson 7: Compare objects using *the same as*.

A STORY OF UNITS

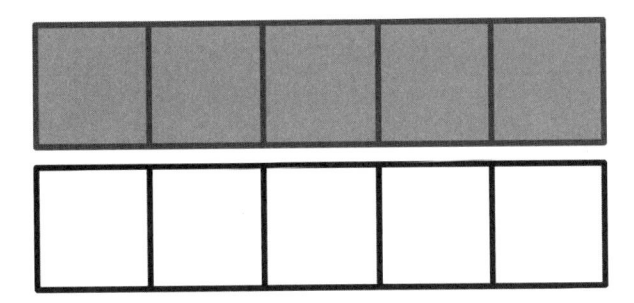

Lesson 7 Homework K•3

Color 1 cube green. Color 4 cubes blue.

How many did you color? _____

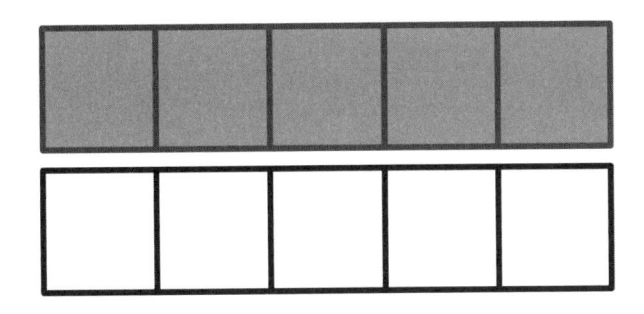

Color 4 cubes green. Color 1 cube blue.

How many did you color? _____

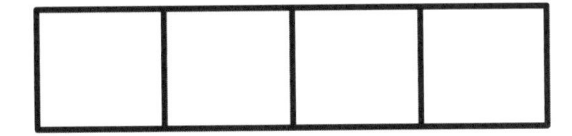

Color 2 cubes yellow. Color 2 cubes blue.

Together, my 2 yellow and 2 blue are the same as _____.

Lesson 7:   Compare objects using *the same as*.

My 5:

My ____:

My ____:

riddle work mat

A STORY OF UNITS

GRADE K

Mathematics Curriculum

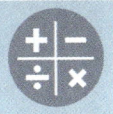

GRADE K • MODULE 3

Topic C
Comparison of Weight

K.MD.1, K.MD.2

Focus Standards:	K.MD.1	Describe measurable attributes of objects, such as length or weight. Describe several measurable attributes of a single object.
	K.MD.2	Directly compare two objects with a measurable attribute in common, to see which object has "more of"/"less of" the attribute, and describe the difference. For example, directly compare the heights of two children and describe one child as taller/shorter.
Instructional Days:	5	
Coherence -Links from:	GPK–M4	Comparison of Length, Weight, Capacity, and Numbers to 5
-Links to:	G1–M3	Ordering and Comparing Length Measurements as Numbers

In Topics A and B, students compared length and height; now, in Topic C, they compare the weight of objects, progressing from informal comparisons of objects (comparing the weight of a book to that of a pencil by picking them up) to using balance scales when greater precision is necessary or desired.

In Lesson 8, students compare the weight of a book to the weight of an eraser and other objects they find. Students then use the weight of the book as a benchmark and find other objects to compare with the weight of the book. "This eraser is lighter than my book. The bag of blocks is heavier than my book."

In Lesson 9, students use a balance scale as a tool to compare the weights of objects that are approximately the same and thus more difficult to compare. For example, "My pencil is lighter than this marker."

In Lesson 10, the measurement becomes more precise as a set of pennies is used to directly compare the weight of objects. Students use a balance to determine that the pencil weighs the same as 5 pennies. The marker weighs the same as 9 pennies. The students compare one object to another, a set and a solid object. They stay within kindergarten standards by not comparing the number of pennies each object weighs; instead they simply enjoy the exploration of finding the set of pennies that weighs as much as an object.

In Lesson 11, students observe conservation of weight. They place, for example, two balls of clay of equal weight on either side of a balance scale. They break one of the balls into two smaller balls and observe the two sides of the scale are still balanced. Students then break the single ball into three smaller balls and observe the same thing. The lesson continues with a sequence leading back to the two balls once again balancing after all the permutations.

In Lesson 12, they extend their learning to use different units to compare the weight of the same item using different objects. "The pencil weighs the same as a set of 5 pennies. The pencil weighs the same as a set of 10 little cubes."

A STORY OF UNITS

Topic C

K•3

A Teaching Sequence Toward Mastery of Comparison of Weight

Objective 1: Compare using *heavier than* and *lighter than* with classroom objects.
(Lesson 8)

Objective 2: Compare objects using *heavier than*, *lighter than*, and *the same as* with balance scales.
(Lesson 9)

Objective 3: Compare the weight of an object to a set of unit weights on a balance scale.
(Lesson 10)

Objective 4: Observe conservation of weight on the balance scale.
(Lesson 11)

Objective 5: Compare the weight of an object with sets of different objects on a balance scale.
(Lesson 12)

A STORY OF UNITS — Lesson 8 K•3

Lesson 8

Objective: Compare using *heavier than* and *lighter than* with classroom objects.

Suggested Lesson Structure

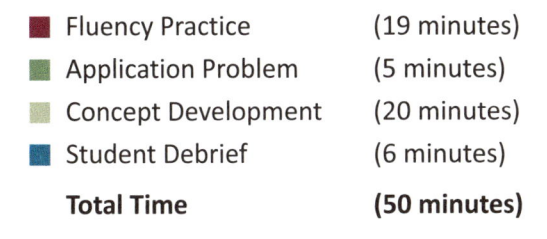

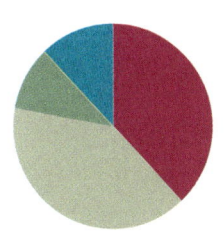

- **Fluency Practice** (19 minutes)
- **Application Problem** (5 minutes)
- **Concept Development** (20 minutes)
- **Student Debrief** (6 minutes)
- **Total Time** **(50 minutes)**

Fluency Practice (19 minutes)

- Make It Equal **K.CC.6** (6 minutes)
- Counting the Say Ten Way with the Rekenrek **K.NBT.1** (4 minutes)
- Beep Number **K.CC.4a** (4 minutes)
- Draw More or Cross Out to Make 5 **K.OA.3** (5 minutes)

Make It Equal (6 minutes)

Materials: (S) Bag of beans, laminated paper or foam work mat, 2 dice

Note: In this activity, students experience comparison visually, a skill crucial to the work of this module.

1. The teacher introduces the term *equal* as meaning *the same number*.
2. Both partners roll the dice and then put that many beans on their mat.
3. Partner A has to make her beans equal to her partner's by taking off or putting on more beans.
4. Partner B counts to verify.
5. Switch roles and play again.

Variation: Have students line up their beans (up to 10 beans) in horizontal or vertical rows.

Counting the Say Ten Way with the Rekenrek (4 minutes)

Conduct activity as outlined in Lesson 6, but now continue to 20 (2 ten) if students are ready.

Note: This activity is an extension of students' previous work with the Rekenrek and anticipates working with teen numbers.

Lesson 8: Compare using *heavier than* and *lighter than* with classroom objects.

Beep Number (4 minutes)

Materials (optional): (T) Personal white board (S) Number path (Lesson 1 Fluency Template)

Note: This activity extends students' proficiency with number order and anticipates working with teen numbers.

Conduct the activity as outlined in Kindergarten Module 1 Lesson 15, but with teen number sequences, counting the Say Ten way. Numbers after are easier to determine than numbers before. Crossing ten proves difficult.

A possible sequence, moving from simple to complex, is the following:

> Ten 1, ten 2, beep
> Ten 6, beep, ten 8
> Beep, ten 4, ten 5
> 9, beep, ten 1

Variation: Extend the sequences to four numbers, for example, ten 1, ten 2, beep, ten 4.

Draw More or Cross Out to Make 5 (5 minutes)

Materials: (S) Make 5 (Fluency Template)

After giving clear instructions and completing the first few problems together, allow students time to work independently. Encourage them to do as many problems as they can within a given time frame.

Optionally, review the answers as a class. Direct students to energetically shout "Yes!" for each correct answer.

Application Problem (5 minutes)

Draw three things you would not mind carrying around in your backpack, even if you had to walk a long way.

Now, draw one thing that you would not want to carry around in your backpack because it might make you very tired. Why wouldn't you want to carry it? How is it different from the first things you drew? Talk to your partner about your pictures.

Note: Thinking about things that are too heavy to carry provides the anticipatory set for today's lesson about relative weights.

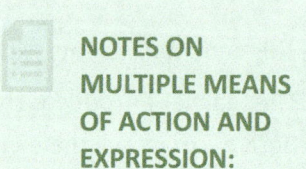

NOTES ON MULTIPLE MEANS OF ACTION AND EXPRESSION:

Connect the directions for the Application Problem with gestures that illustrate their meaning. For instance, when telling students to draw one thing that they would not want to carry around in their backpack because it might make them very tired, the teacher can move slowly and breathe heavily as if out of breath.

| A STORY OF UNITS | Lesson 8 | K•3 |

Concept Development (20 minutes)

Collect objects from the classroom of varying weights, enough that each pair of students has at least three objects to test. Include something tall but light (such as a bag of rice cakes) and something short but heavy (such as a can). Other suggestions include: a stack of books, a pencil, an eraser, a marker, a balloon, a tower of linking cubes, a block, a sphere, some cotton balls, some rocks, and a bag of coins. Be sure to include some surprises that are large but relatively light and some that are small but relatively heavy! Place the objects on a table in the front of the room prior to the beginning of the lesson.

MP.2

T: Look at the rice cakes and the can. Which is taller?
S: The bag of rice cakes.
T: Which would you rather carry home from the store?
S: The bag of rice cakes!
T: Why?
S: It is lighter.
T: We've been talking about how tall, short, or long things are. There are other ways we can compare things, aren't there? Can you think of some more?
S: (Various responses.) We could compare two shoes, which is longer and which is heavier.
→ We could compare a dog and a cat, which is lighter but which is bigger.
T: Good thinking. Let's do some comparing right now. Student A, would you come to the front and be my helper? Please pick up this book.
S: (Pick up the book.)
T: Thank you. Now, pick up these cotton balls. (Student A picks them up.) Which would you rather carry in your backpack all day?
S: The cotton balls!
T: Why?
S: They are lighter.
T: The cotton balls are **lighter than** the book. The book is **heavier than** the cotton balls. We can say that they have different weights. **Weight** is the math word for how heavy or light something is. Thank you, Student A.
T: Student B, would you please come up? Hold the book in one hand and the rocks in the other. What do you notice?
S: The rocks are heavier. → They pull this hand to the floor more. → The book is lighter.
T: The rocks are heavier than the book. The book is lighter than the rocks. Compare using the word *than*.
S: The rocks are heavier *than* the book.
T: They have different weights. Put down the book, and find something else that is lighter than the rocks. (Allow the student to choose another object. Discuss the student's choice, and ask how he or she determined that the object was lighter.)

NOTES ON MULTIPLE MEANS OF ENGAGEMENT:

Scaffold the lesson for students working below grade level by having them practice the concept using interactive technology such as the one found at http://www.ixl.com/math/kindergarten/light-and-heavy.

Lesson 8: Compare using *heavier than* and *lighter than* with classroom objects.

T: Thank you! Student C, do you think the book will be heavier than the eraser or lighter than the eraser? (Allow various students to predict and test the relative weights of various objects against the weight of the book. Discuss how the students determined their answers.)

T: I am going to give you and your partner each some objects. First, make some guesses, and then hold each of them to feel its weight. Work together to see which of your things is the lightest and which is the heaviest. (Allow time for experimentation and discussion. Circulate to encourage use of correct *heavier than* and *lighter than* vocabulary.)

T: Which object did you find to be the heaviest in your group? Hold it up! Which was the lightest? Hold it up. Are the biggest things always the heaviest? (Allow time for discussion.)

Problem Set (10 minutes)

Students should do their personal best to complete the Problem Set within the allotted time.

Student Debrief (6 minutes)

Lesson Objective: Compare using *heavier than* and *lighter than* with classroom objects.

The Student Debrief is intended to invite reflection and active processing of the total lesson experience.

Invite students to review their solutions for the Problem Set. They should check work by comparing answers with a partner before going over answers as a class. Look for misconceptions or misunderstandings that can be addressed in the Debrief. Guide students in a conversation to debrief the Problem Set and process the lesson.

Any combination of the questions below may be used to lead the discussion.

- What did you notice about **heavier than** and **lighter than** when you were working with your partner on our class activity?
- How could you tell that one thing was lighter than or heavier than another?
- Are larger objects always heavier than smaller objects?
- Are smaller objects always lighter than larger objects?
- How did you decide which objects on the Problem Set would be heavier? Could you make a prediction even though you couldn't feel their **weight**?
- Which objects did you circle? Why?
- What new (or significant) math vocabulary did we use today to talk about our object?

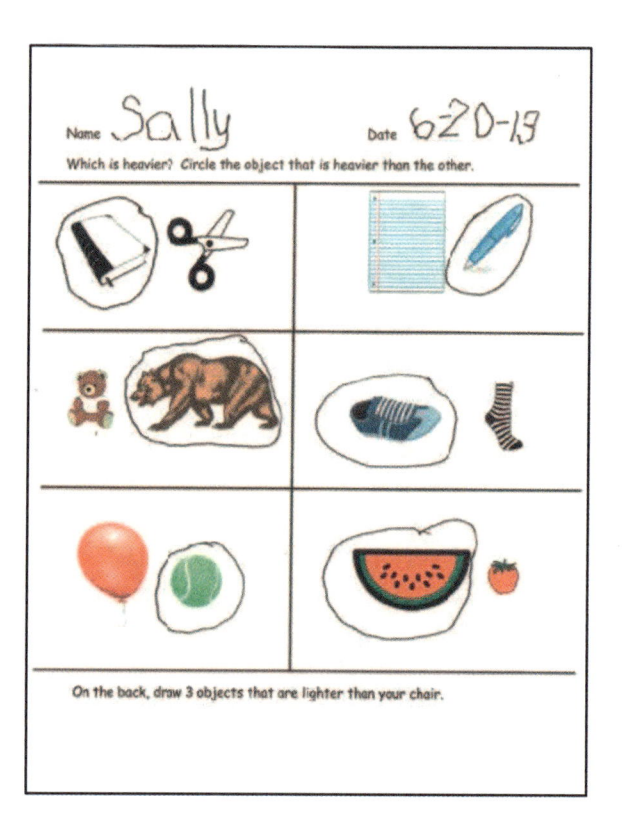

A STORY OF UNITS										Lesson 8 Problem Set K•3

Name _____   Date _____

Which is heavier? Circle the object that is heavier than the other.

On the back, draw 3 objects that are lighter than your chair.

Lesson 8:   Compare using *heavier than* and *lighter than* with classroom objects.

83

A STORY OF UNITS

Lesson 8 Homework K•3

Name _____ Date _____

Draw an object that would be lighter than the one in the picture.

Lesson 8: Compare using *heavier than* and *lighter than* with classroom objects.

Draw more objects, or cross out objects to make 5. Circle the group of 5.

make 5

Lesson 9

Objective: Compare objects using *heavier than, lighter than,* and *the same as* with balance scales.

Suggested Lesson Structure

- ■ Fluency Practice   (14 minutes)
- ■ Application Problem   (5 minutes)
- ■ Concept Development   (25 minutes)
- ■ Student Debrief   (6 minutes)
  - **Total Time**   **(50 minutes)**

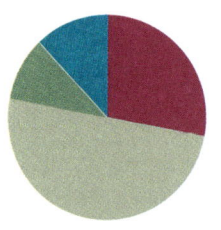

Fluency Practice (14 minutes)

- Hidden Numbers **K.OA.3**   (5 minutes)
- 5-Group Hands **K.CC.2**   (4 minutes)
- Roll and Draw 5-Groups **K.OA.3** (5 minutes)

Hidden Numbers (5 minutes)

Materials: (S) Hidden numbers mat (Lesson 3 Fluency Template) inserted into personal white board

Note: Conduct the activity as described in Lesson 3; however, this time, guide students to find hidden numbers within a group of 8.

5-Group Hands (4 minutes)

Materials: (T) Large 5-group cards (5–7) (Lesson 5 Fluency Template 1)

Note: This maintenance activity develops flexibility in seeing the 5-groups vertically or horizontally and adds a kinesthetic component.

> T: (Show the 6-dot card.) Raise your hand when you know how many dots are on top. (Wait until all hands are raised, and then signal.) Ready?
> S: 5.
> T: Bottom?
> S: 1.

A student demonstrates 7 as 5 on top and 2 on the bottom.

A STORY OF UNITS   Lesson 9 K•3

T: We can show this 5-group on our hands. 5 on top: 1 on the bottom, like this. (Demonstrate on hands, one above the other.)
S: (Show 5 and 1 on hands, one above the other.)
T: Push your hands out as you count on from 5, like this: 5 (extend the top hand forward), 6 (extend the bottom hand forward). Try it with me.
S: 5 (extend the top hand forward), 6 (extend the bottom hand forward).

Continue with 5, 6, 7, steadily decreasing guidance from the teacher, until students can show the 5-groups on their hands with ease.

Roll and Draw 5-Groups (5 minutes)

Materials: (S) Die (with the 6-dot side covered), personal white board

Note: Observe to see which students erase completely each time and begin with one rather than draw more or erase some to adjust to the new number. By drawing 5-groups, students see numbers as having length in relationship to the five.

Conduct the activity as outlined in Lesson 7.

Application Problem (5 minutes)

Put the following sentence frame on the board, and then read it to the students.

   I am lighter than _____, but I am heavier than _____.

Draw two things on your paper that would make this sentence true for you. Show your pictures to your partner. Does he or she agree with you? How much do you think you weigh?

Note: This problem bridges the relative weight comparisons in yesterday's exercise to today's more precise focus using a balance. The balance scale helps students recall times when they themselves have been weighed, for example, at the doctor's office. It also allows the teacher to see what general perceptions the students have about the measurement of weight.

> **NOTES ON MULTIPLE MEANS OF REPRESENTATION:**
>
> Scaffold the directions for English language learners by using gestures while reading the sentence one section at a time: "I am lighter" with a slow lift from the hand, and "I am heavier" with a quick drop of the hand. While gesturing, hold something light for the "I am lighter" sentence and something heavy for the "I am heavier" sentence.

Lesson 9:   Compare objects using *heavier than*, *lighter than*, and *the same as* with balance scales.   87

Concept Development (25 minutes)

Materials: (T) Lighter or heavier recording sheet (Template) affixed to the white board (S) Simple balance scale and assortment of objects such that each small group of students has at least three things to compare (include some objects that are the same weight); lighter or heavier recording sheet (Template)

T: Sometimes when we are comparing the weights of things that are almost the same, it is hard to tell which is lighter and which is heavier. Can you give me an example from yesterday? Was it sometimes hard to tell which thing was heavier?

S: When we compared the marker and the eraser. → It was hard to tell with the balloon and the cotton ball.

T: We have a special tool that can help us find out which object is lighter and which is heavier or if they are the same weight. It is called a **balance scale** or a **balance**. (Display the balance scale. Ask students what they know about the balance.)

T: If I were to put the cotton balls on this side (point to one side of the balance) and the eraser on the other side (point), what would happen?

S: The eraser side would go down. → The side that is heavier will be lower.

T: Let's test your guess. (Demonstrate.) You were right! The balance scale shows us that the eraser is heavier than the cotton balls. It shows us that the cotton balls are lighter than the eraser. I will draw the cotton balls and the eraser in the right places on the lighter or heavier recording sheet. (Demonstrate.)

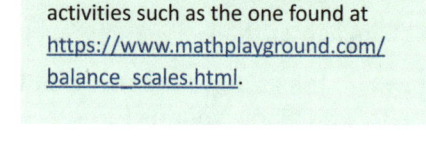

T: Repeat with other pairs of objects until the students are comfortable with the technique of predicting and then experimenting. Draw each pair of items on the lighter or heavier recording sheet.

T: In your small groups, you will be comparing the weights of several pairs of things. You will take turns.

1. Student A chooses two things to compare.
2. Test them first by just holding them and silently guessing which will be heavier.
3. Pass them around so your friends get a chance to guess, too!
4. Student A puts one object on one side of the balance and the other object on the other side of the balance to test the guesses.
5. All of you will record the results on your own lighter or heavier recording sheet.
6. Then, it will be the next student's turn to choose. (Allow ample time for experimentation and recording. Circulate to ensure accurate use of the materials and recording of the results.)

T: Put your balances away. What did your group discover? Were there any surprises? Did anyone find some objects that were the same weight? How did you know? (Allow time for discussion.)

> **NOTES ON MULTIPLE MEANS OF ACTION AND REPRESENTATION:**
>
> Extend the understanding of heavier than, lighter than, and equal to by introducing students working above grade level to interactive balance scale activities such as the one found at https://www.mathplayground.com/balance_scales.html.

Problem Set (10 minutes)

In this lesson, the Problem Set is replaced with the *lighter or heavier recording sheet* to be used during the Concept Development.

Student Debrief (6 minutes)

Lesson Objective: Compare objects using *heavier than, lighter than,* and *the same as* with balance scales.

The Student Debrief is intended to invite reflection and active processing of the total lesson experience.

Invite students to review their lighter or heavier recording sheet. They should check work by comparing answers with a partner before going over answers as a class. Look for misconceptions or misunderstandings that can be addressed in the Debrief. Guide students in a conversation to debrief the *lighter or heavier recording sheet* and process the lesson.

Any combination of the questions below may be used to lead the discussion.

- Why is a **balance scale** helpful?
- Which objects did you record as heavier than? Which ones were lighter than?
- Did you find any objects that were about the same weight?
- Were you surprised by anything you discovered in the activity?
- Explain to your friend which objects you recorded as being lighter or heavier. Did you have the same answer as your friend?

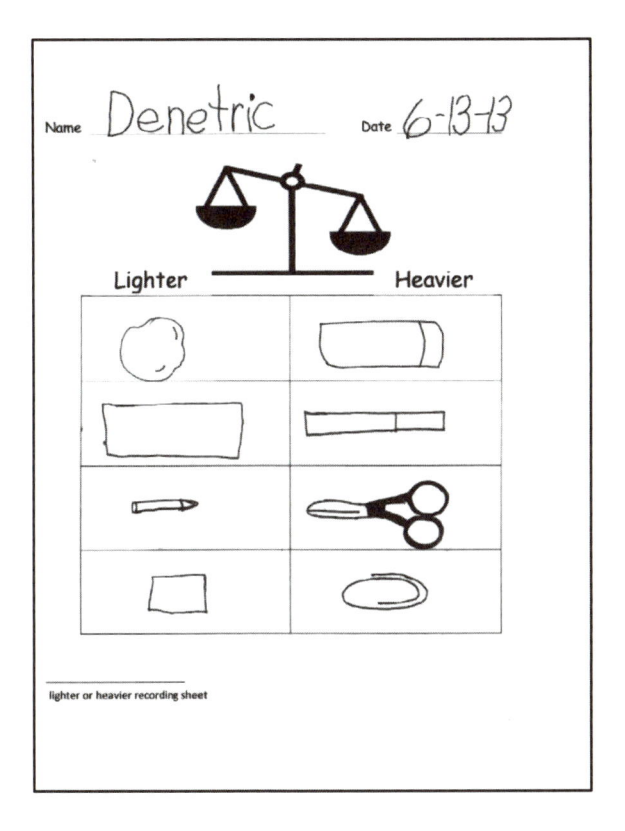

Name _____ Date _____

Draw something inside the box that is heavier than the object on the balance.

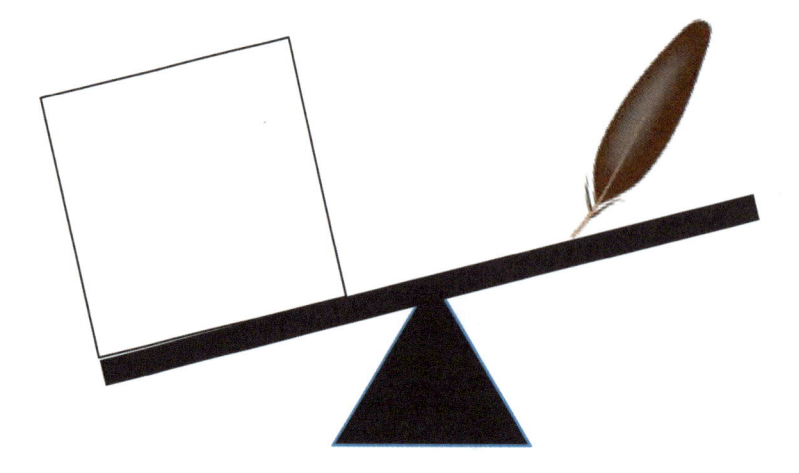

Lesson 9 Homework

Draw something lighter than the object on the balance.

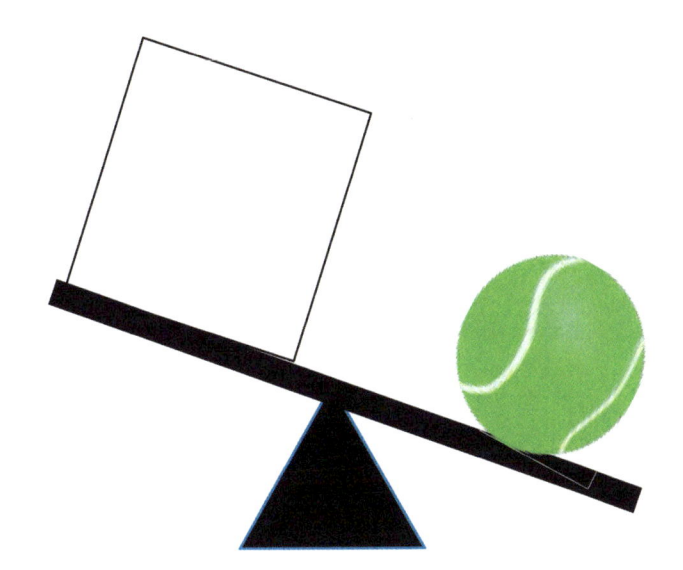

Name _____   Date _____

Lighter   Heavier

lighter or heavier recording sheet

Lesson 10

Objective: Compare the weight of an object to a set of unit weights on a balance scale.

Suggested Lesson Structure

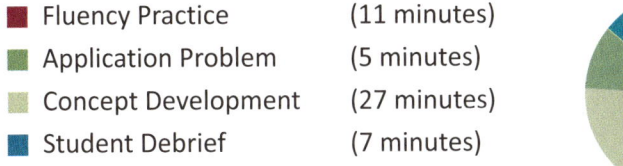

- ■ Fluency Practice   (11 minutes)
- ■ Application Problem   (5 minutes)
- ■ Concept Development   (27 minutes)
- ■ Student Debrief   (7 minutes)
-   **Total Time**   **(50 minutes)**

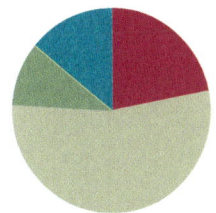

Fluency Practice (11 minutes)

- Green Light, Red Light **K.CC.2**   (3 minutes)
- Make It Equal **K.CC.6**   (4 minutes)
- Double 5-Groups **K.CC.2**   (4 minutes)

Green Light, Red Light (3 minutes)

Materials: (T) Green and red dry-erase markers

Conduct activity as described in Lesson 7, gradually building up to teen numbers counting the Say Ten way. Listen carefully for hesitation or errors, and repeat and break down certain sequences as needed.

Make It Equal (4 minutes)

Materials: (S) Bag of beans, foam or laminated paper work mat, 2 dice

Note: In this activity, students experience comparison visually, a skill crucial to the work of this module.

1. Teacher introduces the term *equal* as meaning *the same number*.
2. Both partners roll the dice and put that many beans on their mat.
3. Partner A has to make his or her beans equal to his or her partner's by taking off or putting on more beans.
4. Partner B counts to verify.
5. Switch roles and play again.

Double 5-Groups (4 minutes)

Materials: (T) Large 5-group cards (Lesson 5 Fluency Template 1)

Note: Introducing Say Ten counting now lays the foundation for later work with decomposing teen numbers.

T: You're getting so good at 5-groups! Now, we'll start using two cards! (Display the 10-dot card above the 1-dot card.) This is the top card. (Gesture to indicate the entire 10-dot card, not just the top row of dots.) How many dots are on the top card? (Wait for all hands to go up, and then give the signal.) Ready?

S: 10.

T: This is the bottom card. (Gesture to indicate the entire 1-dot card.) How many dots are on the bottom card? (Wait for all hands to go up, and then give the signal.) Ready?

S: 1.

T: Do you remember how many dots were on the top card?

S: Yes. 10.

T: Do we really need to go back and count them again?

S: No.

T: That's right. We can take the shortcut! Count on from 10, like this. 10 (Wave a hand over the top card.) Ten 1. (Crisply point to the dot on the bottom card.) Try it.

S: 10, ten 1.

T: (Display the 10-dot card above the 2-dot card.) How many dots are on the top card? (Wait for all hands to go up, and then give the signal.) Ready?

S: 10.

T: How many dots are on the bottom card? (Wait for all hands to go up, and then give the signal.) Ready?

S: 2.

T: Count on from 10.

S: 10, ten 1, ten 2.

Continue to ten 3.

Application Problem (5 minutes)

Imagine that you were on a seesaw with a little kitten on the other end. Draw a picture of yourself and the kitten on the seesaw. Which end of the seesaw would be closer to the ground? How do you know? Talk about your picture with your partner. Do your seesaws look the same?

Note: This problem provides students with an opportunity to think about a practical application of a *balance* and to represent it and explain it to their friends. Listen for phrases such as *heavier than* and *lighter than*, and encourage precision in the discussion. The activity bridges the *heavier than* and *lighter than* emphasis from Lesson 9's balance activity with today's more precise use of the tool.

Concept Development (27 minutes)

Materials: (T) Balance scale, pencil, marker, bag of 30 pennies, as heavy as recording sheet (Template) affixed to the white board (S) Balance scale, bag of 30 pennies, bag of objects to weigh (including a pencil, an eraser, a marker, a small child's pair of scissors, a linking cube, and a small block or toy) per pair or small group; as heavy as recording sheet (Template)

T: I have nothing on my balance. What do you notice?
S: It is even. → It's straight across. → It looks the same on both sides.
T: (Place a pencil on one side and a marker on the other side of the balance.) Which is heavier, this pencil or this marker? How do you know?
S: The marker. → The side with the marker is lower.
T: (Remove the marker, and replace it with the eraser.) Which is heavier, the pencil or the eraser?
S: The eraser! That side is lower.
T: I want to find something that is the same weight as my pencil. How would I know if it were the same weight? How would my balance look?
S: It would be the same on both sides. → It would be even!
T: Yes, I would know something weighed the same as the pencil if the balance looked *even*. It would look like this. (Demonstrate. If there is an equilibrium marker on the balance, use this opportunity to show the students how to use the marker.)
T: (Remove the eraser, and replace it with a penny.) Which is heavier, the pencil or the penny?
S: The pencil.
T: (Add another penny.) Which is heavier, the pencil or two pennies?
S: The pencil is still heavier than two pennies!
T: (Continue adding pennies, one at a time, until balanced.)
S: It is even! → They are the same!
T: Let's count the pennies on our balance again.
S: 1, 2, 3, 4, 5. (Answers may vary.)
T: The pencil is as heavy as a set of five of the pennies! I'm going to show that on my recording sheet. (Demonstrate.) Student A, would you please come up to help me test something else? (Empty the balance, and place the marker on one side.)
T: I wonder how many pennies are as heavy as the marker? (Various responses.) Student A, will you help find out? Count with Student A.
S: 1, 2, 3, 4, 5, 6.
T: The marker is as heavy as a set of six pennies. I will put that on my recording sheet. (Demonstrate.)

NOTES ON MULTIPLE MEANS OF REPRESENTATION:

In addition to your model, scaffold the lesson for students working below grade level by using pictures of what a balance scale looks like when one object is heavier or lighter than the pennies used on the other side or when the scale is balanced and the objects are the same weight as the pennies. Students can refer to the visual as an aid.

Lesson 10: Compare the weight of an object to a set of unit weights on a balance scale.

A STORY OF UNITS   **Lesson 10**   K•3

MP.6

T: You and your partner are going to compare the weight of pennies with other things in our classroom. Choose one of the objects from your bag. Guess how many pennies will be as heavy as your object. Use your balance to test your guess. On your recording sheet, draw a picture of your object, and then count and write how many pennies weigh the same as your object. (Allow time for experimentation and recording of results.)

T: Put your things away. What did you discover? (Allow time for discussion.) Which object was the heaviest? Which object was the lightest? Were any of them the same weight? (Allow time for discussion.)

Problem Set (10 minutes)

In this lesson, the as heavy as recording sheet serves as the Problem Set for the Concept Development.

Student Debrief (7 minutes)

Lesson Objective: Compare the weight of an object to a set of unit weights on a balance scale.

The Student Debrief is intended to invite reflection and active processing of the total lesson experience.

Invite students to review their Recording Sheets. They should check work by comparing answers with a partner before going over answers as a class. Look for misconceptions or misunderstandings that can be addressed in the Debrief. Guide students in a conversation to process the lesson.

Any combination of the questions below may be used to lead the discussion.

- What did you notice as you weighed the objects?
- When you guessed how many pennies each object would weigh, how close were you?
- How did you know when to stop adding pennies to the balance scale?
- Were you surprised by anything that happened in the activity today?
- Show your as heavy as recording sheet to your friend. Did she make some of the same discoveries?
- What new (or significant) math vocabulary did we use today to communicate precisely?

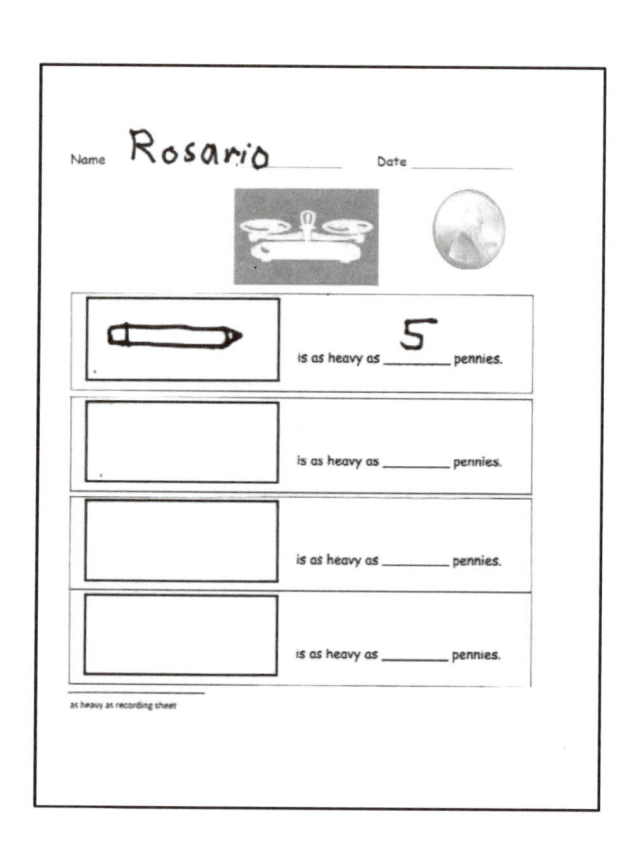

Name _____ Date _____

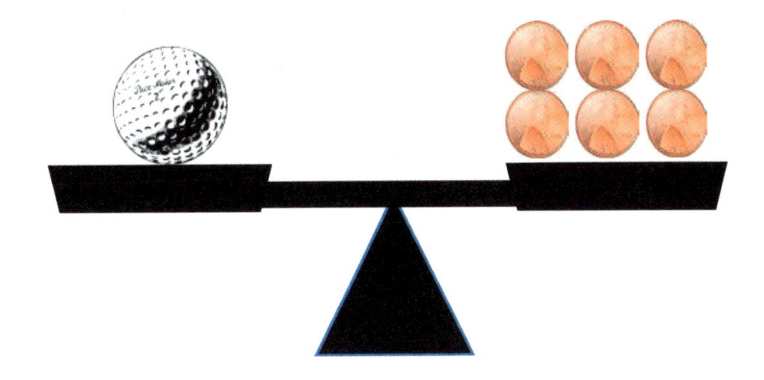

The golf ball is as heavy as _____ pennies.

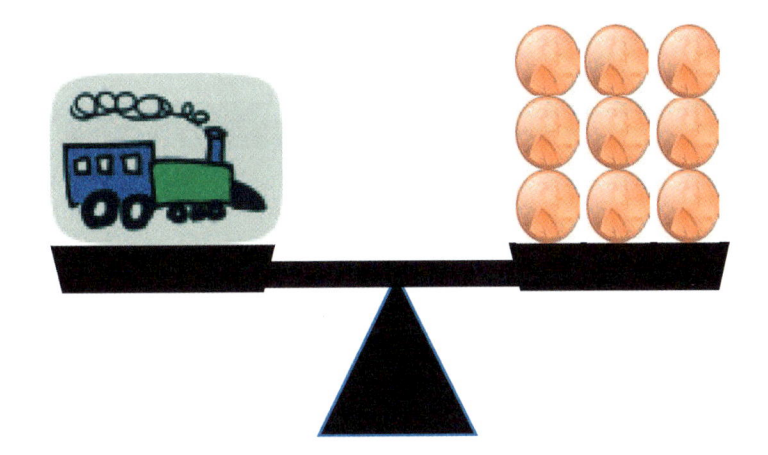

The toy train is as heavy as _____ pennies.

Lesson 10: Compare the weight of an object to a set of unit weights on a balance scale.

A STORY OF UNITS

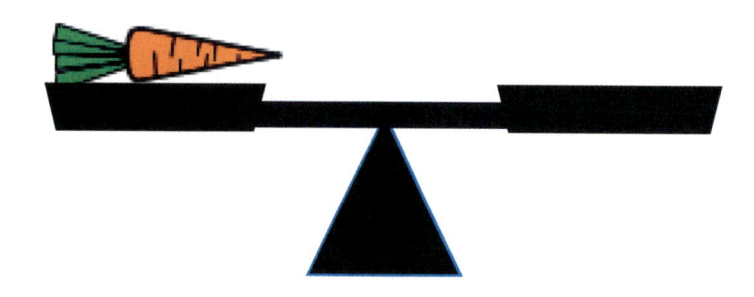

Lesson 10 Homework K•3

Draw in the pennies so the carrot is as heavy as 5 pennies.

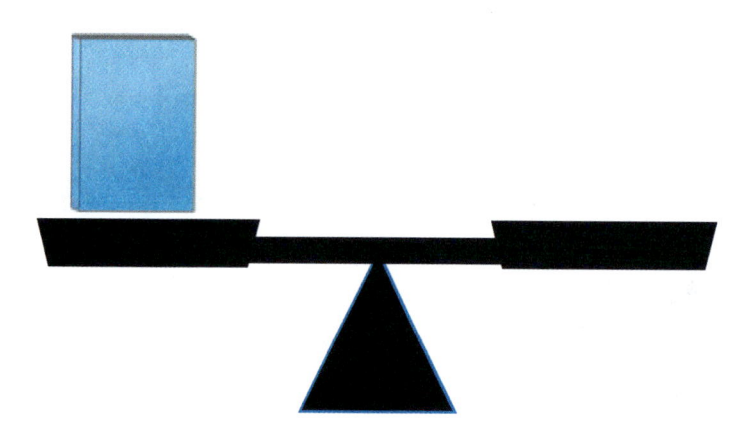

Draw in the pennies so the book is as heavy as 10 pennies.

On the back of your paper, draw a balance scale with an object. Write how many pennies you think the object would weigh. If you can, bring in the object tomorrow. We will weigh it to see if it weighs as many pennies as you thought.

Name _____   Date _____

	is as heavy as _____ pennies.
	is as heavy as _____ pennies.
	is as heavy as _____ pennies.
	is as heavy as _____ pennies.

as heavy as recording sheet

Lesson 10: Compare the weight of an object to a set of unit weights on a balance scale.

Lesson 11

Objective: Observe conservation of weight on the balance scale.

Suggested Lesson Structure

- **Fluency Practice**   (13 minutes)
- **Application Problem** (5 minutes)
- **Concept Development** (25 minutes)
- **Student Debrief**   (7 minutes)
- **Total Time**   **(50 minutes)**

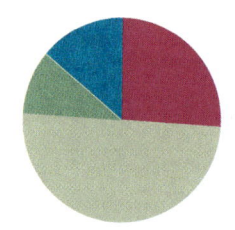

Fluency Practice (13 minutes)

- Heavier or Lighter **K.MD.1**   (4 minutes)
- Double 5-Groups **K.CC.2**   (4 minutes)
- Hidden Numbers **K.OA.3**   (5 minutes)

Heavier or Lighter (4 minutes)

Materials: (T) Balance scale and assorted objects

Note: This activity prepares students for today's lesson by reviewing vocabulary, isolating the attribute of weight, and incorporating a kinesthetic component to enhance conceptual understanding.

- T: Look at my objects. (Show a cotton ball and an orange, for example.) I'm going to put them on the scale. Watch carefully to see how the scale moves. Raise your hand when you know which one is heavier. (Wait for all hands to go up, and then give the signal.) Ready?
- S: The orange!
- T: Yes! Now, pretend you're the scale! Show me the side that is heavier.
- S: (Pretend to hold the orange in one hand, and then quickly lower the hand to indicate weight.)
- T: Now, raise your hand when you know which is lighter. (Wait for all hands to go up, and then give the signal.) Ready?
- S: The cotton ball!
- T: Yes! Now, pretend you're the scale! Show me the side that is lighter.
- S: (Pretend to hold the cotton ball in one hand, and gradually lift it as if it is being pulled up by a balloon.)

Continue with a variety of objects, especially those that produce unexpected results. Compare a large feather to a small rock so that students can see that size does not always correlate to weight.

Double 5-Groups (4 minutes)

Materials: (T) Large 5-group cards (Lesson 5 Fluency Template 1)

Note: Introducing Say Ten counting now lays the foundation for later work with decomposing teen numbers. Conduct the activity as outlined in Lesson 10, but now continue to ten 5.

Hidden Numbers (5 minutes)

Materials: (S) Hidden numbers mat (Lesson 3 Fluency Template) inserted into personal white board

Note: Finding embedded numbers anticipates the work of Module 4 by developing part–whole thinking.

Conduct the activity as described in Lesson 3, but this time, guide students to find hidden numbers within a group of 9.

Application Problem (5 minutes)

Materials: (S) Small bag of about 10 Lego-type building blocks, balance scale for small group, 20 pennies

Use your blocks to make the heaviest building that you can. How many pennies are as heavy as your building? Turn to your friend. Talk about your different buildings and how much they weigh.

Note: This question allows students to puzzle over and discuss whether or not the configuration of their building affects its weight, serving as an anticipatory set for today's lesson.

Concept Development (25 minutes)

Materials: (T) Balance scale, ball of clay (S) Balance scale, ball of clay (per small group or pair)

T: I have a ball of clay for each pair of students. When you get your clay, Partner A will make it into two balls that are about the same size.
S: (Pass out clay to Partner A.)
T: Now, here is a balance scale. Partner B, put the two new balls on each side of the scale.
S: (Partner B puts them on the scale.)
T: Talk to your partner. Are the balls the same weight? Use the math words *heavier than* and *lighter than*, please.
S: No. This one is heavier than the other because it went down. → This ball is lighter because the scale is going up on this side.

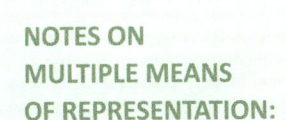

NOTES ON MULTIPLE MEANS OF REPRESENTATION:

Introduce the term *heaviest* to English language learners before the lesson so that they can understand the directions and gain the full benefit of the Application Problem. Show visuals of *heavy, heavier,* and *heaviest* on the word wall and refer to them while giving directions. Have them practice saying, "Object A is *heavier than* Object B," "Object B is *lighter than* object C," etc., in preparation for the lesson.

Lesson 11: Observe conservation of weight on the balance scale.

T: (Show a balance that has a ball of clay on each side, one heavier than the other.) Look at my scale. Point to the ball that is heavier.

S: (Point.)

T: The one that is lighter?

S: (Point.)

T: I want to make the balls the same weight, so I will take a bit of clay from the heavier one and move it to the lighter one. I'll keep moving little pieces until they balance. I'm making sure to keep my clay in the middle, not on the edge.

T: Now, you try. Take turns moving pieces from the heavier ball to the lighter ball until they balance, until they weigh the same amount.

S: (Manipulate the clay until they are equal.)

T: Are your clay balls the same weight now?

S: Yes.

T: Remove one ball, all of you. Without removing any clay, Partner A, make your ball nice and round. Partner B, make your ball into a pancake. You have 30 seconds.

T: Put your clay back on the scale. Do they still weigh the same amount?

S: Yes!

T: Partner B, I want you to take your pancake off and quickly make all of it into two balls. (Pause.)

T: Talk to your partner. What do you think will happen when you put the two smaller balls back on the scale?

S: They are going to weigh more because they are more now. → I think it's going to be the same because we didn't take any off. → I think it's going to weigh less because they are smaller.

T: Okay. Put the balls back on the scale.

S: They are still the same!

T: How are they the same? Are they the same number? The same size?

S: No. The same weight!

T: Let's try another experiment. Partner A, take your ball and quickly make it into three smaller balls.

T: Talk to your partner. What will happen this time when Partner A puts his or her part back on the scale?

S: It's going to balance. → No. This time there is more, so it's going to weigh more. → No. I think it's the same even though there are more pieces. → It balanced before, so it will this time, too. → It would only change if we took off clay and didn't put it back.

T: Put the three balls back on the scale.

T: Are the two sides of the balance showing the same number of balls?

S: No. One side has two balls, and the other has three balls.

T: Are they the same size?

S: No. These ones are smaller.

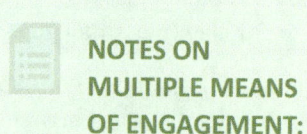

> **NOTES ON MULTIPLE MEANS OF ENGAGEMENT:**
>
> Ask students working above grade level what would happen if you placed both clay balls on one side and placed the building blocks on the other side. Would the two sides of the balance scale be equal? Ask them to explain why they balanced (or did not balance) the scale.

A STORY OF UNITS	Lesson 11 K•3

T: Are they the same weight?

S: Yes, they are!

Continue the process, moving it along as quickly as possible so that students stay focused on the weighing rather than the manipulation of the clay. Consider having Partner A put his two balls together to make one ball and see if it balances Partner B's three smaller balls. Then have Partner B put his three balls back together into one bigger ball so they end up back where they started with two balls that balance on the scale.

Problem Set (10 minutes)

Students should do their personal best to complete the Problem Set within the allotted time.

Student Debrief (7 minutes)

Lesson Objective: Observe conservation of weight on the balance scale.

The Student Debrief is intended to invite reflection and active processing of the total lesson experience.

Invite students to review their solutions for the Problem Set. They should check work by comparing answers with a partner before going over answers as a class. Look for misconceptions or misunderstandings that can be addressed in the Student Debrief. Guide students in a conversation to debrief the Problem Set and process the lesson.

Any combination of the questions below may be used to lead the discussion.

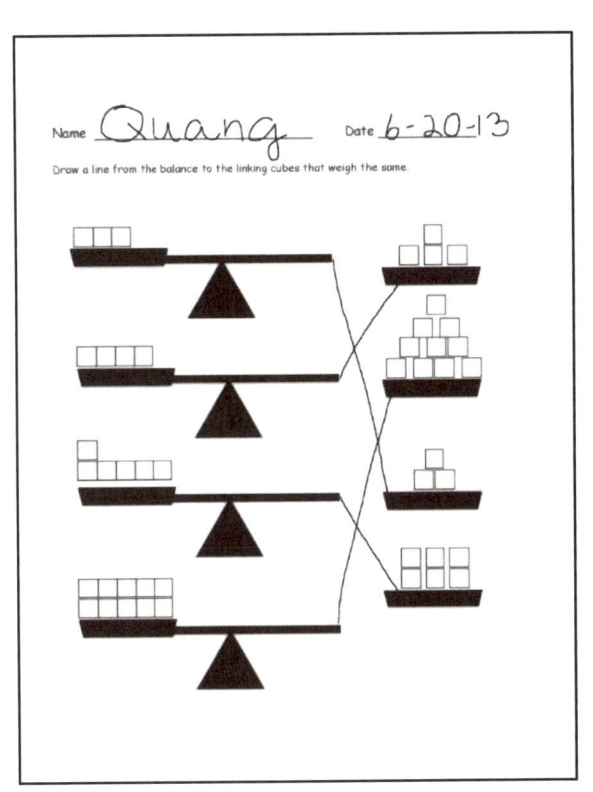

- What happened when you took the clay ball apart, made it into two balls, and weighed them together on the balance?
- What do you think would happen if you took that same clay ball apart, made it into 10 little balls, and put them all on the balance? A hundred little balls?
- Can one thing have the same weight as 10 things? (If you have materials to demonstrate this, all the better. One option is base ten Dienes blocks. In a high-quality set, the thousands cube has the same weight as ten of the hundreds flats.)

Lesson 11: Observe conservation of weight on the balance scale.

A STORY OF UNITS   Lesson 11 Problem Set   K•3

Name _____   Date _____

Draw a line from the balance to the linking cubes that weigh the same.

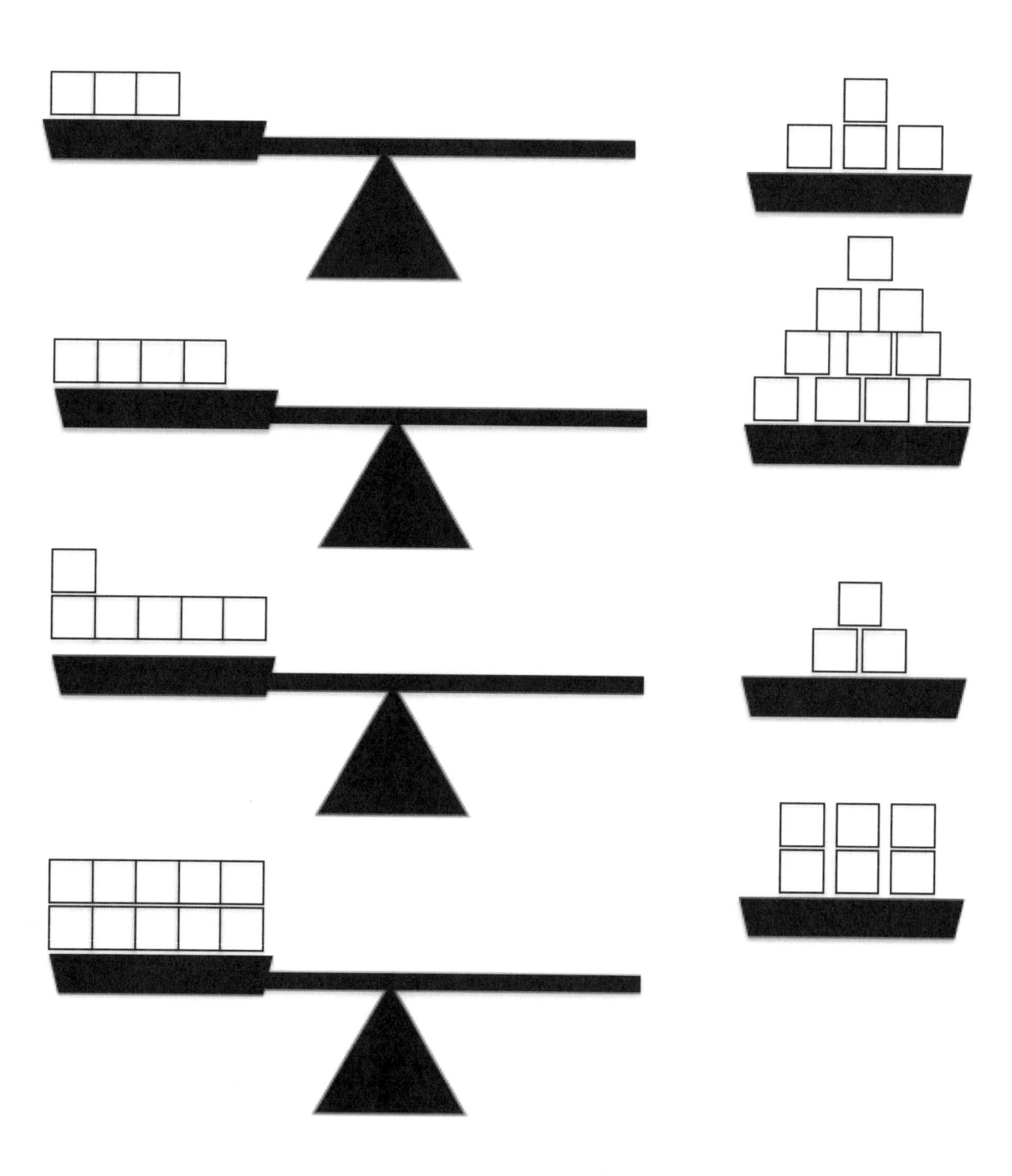

104   Lesson 11:   Observe conservation of weight on the balance scale.

A STORY OF UNITS

Lesson 11 Homework K•3

Name _____ Date _____

Draw linking cubes so each side weighs the same.

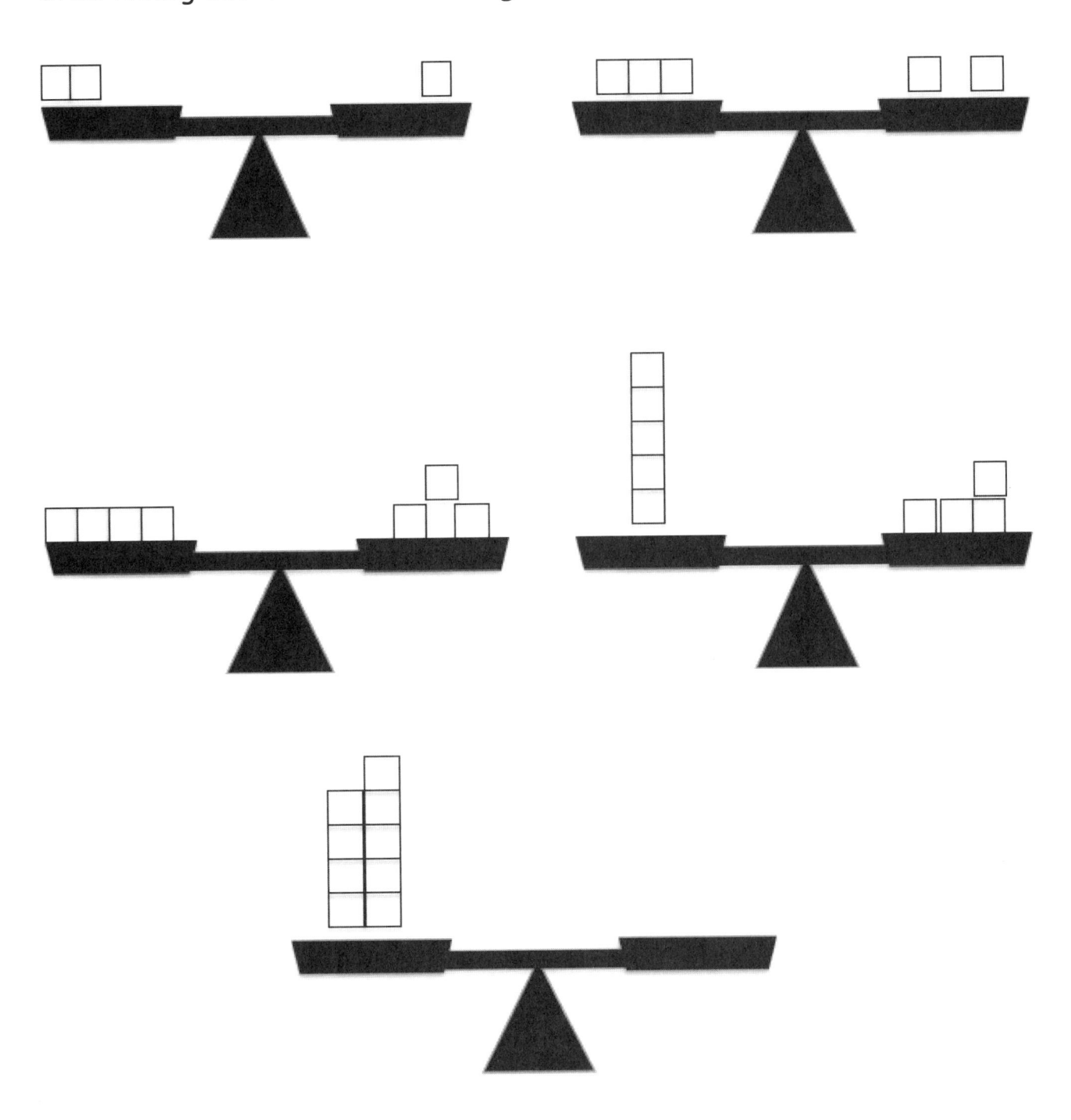

Lesson 12

Objective: Compare the weight of an object with sets of different objects on a balance scale.

Suggested Lesson Structure

- ■ Fluency Practice   (12 minutes)
- ■ Application Problem   (5 minutes)
- ■ Concept Development   (27 minutes)
- ■ Student Debrief   (6 minutes)
  Total Time   **(50 minutes)**

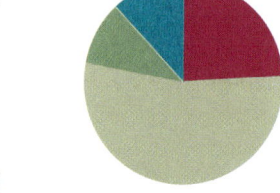

Fluency Practice (12 minutes)

- 5-Group Hands **K.CC.2**   (3 minutes)
- Roll and Draw 5-Groups **K.OA.3**   (5 minutes)
- Hidden Numbers on the Dot Path **K.OA.3**   (4 minutes)

5-Group Hands (3 minutes)

Materials: (T) 5-group cards in vertical orientation (Lesson 5 Fluency Template 1)

Note: This maintenance activity develops flexibility in seeing the 5-groups vertically or horizontally and adds a kinesthetic component.

Conduct as described in Lesson 5, showing the 5-group cards in the vertical orientation. Accordingly, students should put their hands side by side to represent the number.

Roll and Draw 5-Groups (5 minutes)

Materials: (S) Die (with the 6-dot side covered), personal white board

Note: Observe to see which students erase completely and begin each time from one, rather than draw more or erase some to adjust to the new number. By drawing 5-groups, students see numbers as having length in relationship to the five.

Conduct as outlined in Lesson 7. Consider alternating between drawing the 5-groups vertically and drawing them horizontally.

A STORY OF UNITS — Lesson 12 K•3

Hidden Numbers on the Dot Path (4 minutes)

Materials: (S) Dot path (Lesson 5 Fluency Template 2) inserted into personal white board

Note: Finding embedded numbers anticipates the work of Module 4 by developing part–whole thinking.

- T: Fold your dot path so that you can see only 6 dots. Place it inside your personal white board. How many dots can you see?
- S: 6.
- T: Circle 2 of them.
- S: (Circle the first 2 dots.)
- T: See how many twos you can circle on your dot path.
- S: (Circle 3 sets of 2 dots.)
- T: How many dots are on the whole dot path?
- S: 6.
- T: How many twos did you find hiding within the 6?
- S: 3.

Continue the process with finding groups of 3 within the 6. Guide students to find a group of 4 or 5 and then tell what number of dots remains.

Application Problem (5 minutes)

MP.6

Find one small item in your backpack. Put it on the balance scale. How many pennies do you think it will take to balance your object? Use pennies to test your guess. Make a picture of the balance with your object and the pennies. Finish this sentence, "My item is as heavy as a set of _____ pennies."

What do you think would happen if you put another penny on each side of the balance scale? Test your guess!

Note: The review of use of the balance to find objects of equal weight serves as the anticipatory set for today's lesson.

NOTES ON MULTIPLE MEANS OF ENGAGEMENT:

Challenge students working above grade level by asking them to explain why they think the balance scale remains evenly balanced even as they place an extra penny or more on each side.

Concept Development (27 minutes)

Materials: (T) Simple balance scale, marker, 2 pennies, small bag of linking cubes, small counters, beans, and as heavy as a set recording sheet (Template) (S) 1 simple balance scale per pair or small group of students, 4 small bags of various items to use as weights (pennies, linking cubes, small counters, and large dried beans), collection of classroom objects for the balance exercise, and as heavy as a set recording sheet (Template)

- T: Look carefully at my balance. Now, watch as I put my marker on one side. Do you remember how I weighed my marker yesterday?
- S: You used pennies.

Lesson 12: Compare the weight of an object with sets of different objects on a balance scale.

T:	Let's try that again. I have a set of 2 pennies. Watch and see if the scale balances.
S:	1 penny… 2 pennies… It is not enough! The marker is too heavy.
T:	My marker is heavier than a set of 2 pennies. I don't have any more pennies. What should I do?
S:	(Various comments.)
T:	Look at the other items on the table. Is there another way to see how heavy the marker is?
S:	What if we used cubes?
T:	Could I use my two pennies and a cube?
S:	No! That wouldn't be fair. → The cubes and the pennies aren't the same. → You shouldn't count them all together.
T:	I'll take the pennies off and use a tower of cubes. Help me count how many cubes would be in a tower as heavy as my marker.
S:	1, 2, 3, 4, 5, 6. 6 cubes.
T:	My marker is as heavy as a tower of 6 cubes. Let me put that on the recording sheet. I will draw the marker and the cubes, and I will write how many cubes in the box. (Demonstrate.)
T:	What else could I use?
S:	Try the beans!
T:	I will take off the cubes and use a set of beans this time. I wonder how many beans it will take to balance my marker. (Various responses.) Count with me. (Repeat the experiment and recording with beans and small counters.)
T:	Wow! Look what we've discovered. (Point to the sheet.) My marker is as heavy as a tower of 6 cubes. My marker is as heavy as a set of 10 beans. My marker is as heavy as a set of 4 counters. Why are all the numbers different?
S:	The things are all different! → Because the cubes are bigger than the beans. → Because the counters are heavier.
T:	You and your partner can try this, too. Choose one object from your bag. Count how many pennies are as heavy as your object, and record it on your sheet. Then, count how many cubes are as heavy as the object. Do the same thing with the beans and the counters. Don't forget to guess before you test! (Allow time for experimenting and recording. Circulate to make sure that the only variation is in the unit of measurement.)
T:	Put your things away. Who would like to share his or her recording sheet with our class? What did you discover?

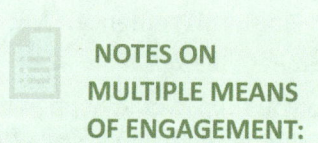

NOTES ON MULTIPLE MEANS OF ENGAGEMENT:

Scaffold the activity for English language learners by providing sentence frames such as, "I think my [object] is [number] pennies heavy." Listen as they use the sentence frames during their partner work, and encourage them.

Problem Set (10 minutes)

In this lesson, the as heavy as a set recording sheet will serve as the Problem Set for the lesson.

Student Debrief (6 minutes)

Lesson Objective: Compare the weight of an object with sets of different objects on a balance scale.

The Student Debrief is intended to invite reflection and active processing of the total lesson experience.

Invite students to review their recording sheets. They should check work by comparing answers with a partner before going over answers as a class. Look for misconceptions or misunderstandings that can be addressed in the Student Debrief. Guide students in a conversation to process the lesson.

Any combination of the questions below may be used to lead the discussion.

- Did you notice any patterns as you were balancing your object with sets of different things?
- Which set of things was the biggest? Which set was the smallest?
- Why were all of the sets different sizes?
- Compare your recording sheet with your friends'. Did you find the same answers?
- What math vocabulary did we use today to communicate precisely?

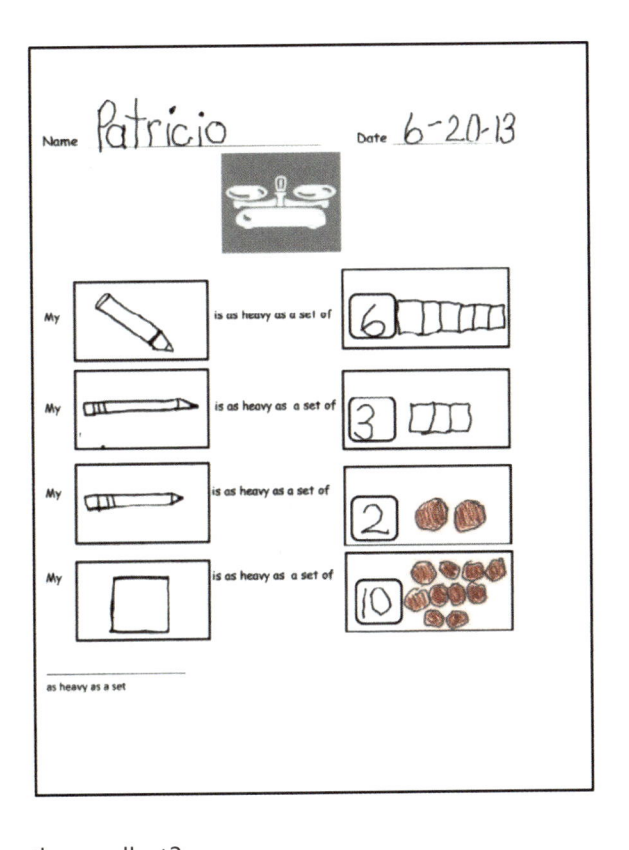

A STORY OF UNITS

Lesson 12 Homework K•3

Name _____   Date _____

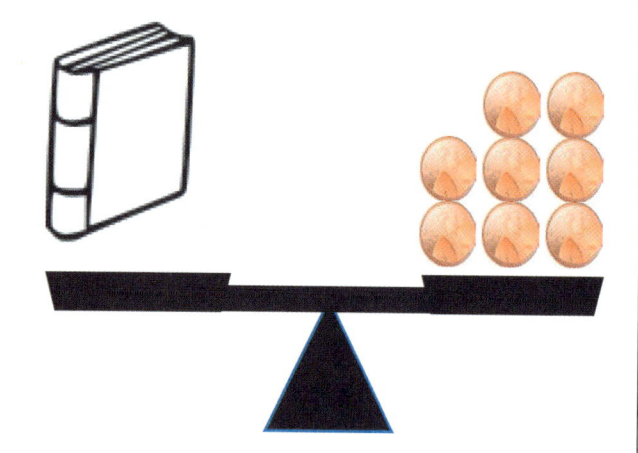

The book is as heavy as _____ pennies.

The book is as heavy as _____ tennis balls.

The book is as heavy as _____ cubes.

The book is as heavy as _____ counting bears.

Lesson 12: Compare the weight of an object with sets of different objects on a balance scale.

A STORY OF UNITS | Lesson 12 Template | K•3

Name _____ Date _____

My [   ] is as heavy as a set of [   ]

My [   ] is as heavy as a set of [   ]

My [   ] is as heavy as a set of [   ]

My [   ] is as heavy as a set of [   ]

as heavy as a set recording sheet

Lesson 12: Compare the weight of an object with sets of different objects on a balance scale.

111

A STORY OF UNITS

Mathematics Curriculum

GRADE K

GRADE K • MODULE 3

Topic D
Comparison of Volume

K.MD.1, K.MD.2

Focus Standards:	K.MD.1	Describe measurable attributes of objects, such as length or weight. Describe several measurable attributes of a single object.
	K.MD.2	Directly compare two objects with a measurable attribute in common, to see which object has "more of"/"less of" the attribute, and describe the difference. *For example, directly compare the heights of two children and describe one child as taller/shorter.*
Instructional Days:	3	
Coherence -Links from:	GPK–M4	Comparison of Length, Weight, Capacity, and Numbers to 5
-Links to:	G1–M3	Ordering and Comparing Length Measurements as Numbers

In Topic D, students compare volume in the same progression as that of weight in Topic C. In Lesson 13, they see that one container holds more rice than another by pouring the rice from the first container into a smaller empty one. "It is overflowing! The bowl holds more rice than the cup."

In Lesson 14, students explore how volume is conserved by pouring rice from a bowl to a bottle and then back into the original bowl. They discover that while the quantity of rice may look very different when poured into containers of different sizes and shapes, the amount remains the same.

In Lesson 15, students count the number of small scoops of rice within a larger amount. "The bowl holds 10 little scoops of rice. I wonder how many little scoops of rice this mug holds?" Before the Mid-Module Assessment, students consider the different measurable attributes of single items such as a water bottle, dropper, and juice box. They consider what tools they might use to compare these attributes.

A Teaching Sequence Toward Mastery of Comparison of Volume
Objective 1: Compare volume using *more than*, *less than*, and *the same as* by pouring. (Lesson 13)
Objective 2: Explore conservation of volume by pouring. (Lesson 14)
Objective 3: Compare using *the same as* with units. (Lesson 15)

Lesson 13

Objective: Compare volume using *more than*, *less than*, and *the same as* by pouring.

Suggested Lesson Structure

- ■ Fluency Practice   (10 minutes)
- ■ Application Problem (5 minutes)
- ■ Concept Development (29 minutes)
- ■ Student Debrief   (6 minutes)
- **Total Time**   **(50 minutes)**

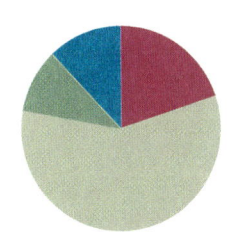

Fluency Practice (10 minutes)

- Dot Cards of 6 **K.CC.2**   (3 minutes)
- Building *1 More* and *1 Less* Towers **K.CC.4c** (4 minutes)
- Roll and Say 1 More, 1 Less **K.CC.4c**   (3 minutes)

Dot Cards of 6 (3 minutes)

Materials: (T/S) Dot cards of 6 (Fluency Template)

Note: This activity deepens students' knowledge of embedded numbers and develops part–whole thinking, foundational to the work of the upcoming modules.

- T: (Show the card.) How many do you see?
- S: 6.
- T: How did you see them in two parts?
- S: (Possible answers are 5 up and 1 down, 2 down and 4 up, 3 up and 3 down.)

Continue with other cards of 6. Distribute the cards to the students for partner sharing time. Have them pass the card at a signal.

> **NOTES ON MULTIPLE MEANS OF REPRESENTATION:**
>
> The dot cards can be adjusted to facilitate the recognition of hidden numbers. Use dots in more than one color, or include some dots with only an outline. This can be used as a modification for students with visual discrimination difficulties, especially during partner sharing time.

Building *1 More* and *1 Less* Towers (4 minutes)

Materials: (S) 10 linking cubes

Note: In this activity, students connect increasing length and height to increasing numerical value.

Guide students through the process of building a tower while stating the pattern as *1 more*. Maintain consistency in the language: 1. 1 more is 2. 2. 1 more is 3. 3. 1 more is 4. (Continue to 10.)

Disassemble the tower while stating the pattern as *1 less*. Again, the language is crucial to students' conceptual understanding: 10. 1 less is 9. 9. 1 less is 8. 8. 1 less is 7. (Continue to 0.)

Consider having students build the towers vertically as towers but also horizontally as a train of cubes.

Roll and Say 1 More, 1 Less (3 minutes)

Materials: (S) Pair of dice with the 6-dot side covered with a sticker

Note: This exercise prepares students for today's lesson by moving flexibly between the terms *more* and *less*.

Roll the dice, and count the dots. Make *1 more* and *1 less* statements using consistent language. For example, if the student rolls a 4, they would say, "4. 1 more is 5. 4. 1 less is 3."

Application Problem (5 minutes)

Materials: (S) Small ball of clay

With your clay, create a cup that could hold just enough milk for a little kitten to drink. Show your cup to your friend. Do you think your cups would hold the same amount?

Note: Thinking about *holding enough* serves as an anticipatory set for the discussion in today's lesson.

Concept Development (29 minutes)

> **NOTES ON MULTIPLE MEANS OF REPRESENTATION:**
>
> Scaffold the Application Problem for English language learners by emphasizing the critical concept of *just enough*. Show visuals of what, for a little kitten, would be *too much* milk (a gallon container), *too little* (a couple of drops), and *just right* (a picture of a cat drinking milk out of a regular-size container).

Materials: (T/S) 2 cups of uncooked rice, several small containers (two with equal capacity: coffee or beverage scoop, ¼ cup measure, teacup, bowl, small drinking cup, small box, tablespoon), and tray per pair or small group; capacity recording sheet (Template)

T: What do you notice on your tray?
S: We have lots of cups. → We have a box. → We have a spoon. → There is a bowl of rice.
T: Watch as I fill my cup with the rice. Tell me when it is full. (Place a medium-sized cup on the tray to prevent any spills. Use the tablespoon to fill it with rice.)
T: How could you tell my cup was full?
S: It was all the way to the top! → No more would fit without spilling.

A STORY OF UNITS Lesson 13 K•3

MP.2

T: It held a lot of rice. One math word for how much something holds is **capacity**. (Hold up a smaller container.) I wonder if the capacity of this container is **more than** or **less than** the capacity of my cup? Do you think it will hold more or less?

S: Less!

T: Repeat after me. "I think the capacity of this container is less than the capacity of the cup."

S: (Repeat.)

T: Let's test your guess. (Pour rice into the smaller cup until it begins to overflow.) What happened?

S: There was too much in the little cup! It spilled! → There wasn't enough room.

T: It was too small. The capacity of the little cup is less than the capacity of the first cup. (Hold up a larger bowl.) Do you think the capacity of this container is more or less than the capacity of my little cup?

S: More! → It has more capacity.

T: Let's test your guess. (Pour rice from the small cup into the larger bowl.) What happened?

S: There is a lot of room left. → We could put more in.

T: The capacity of the bowl is more than the capacity of the little cup! I'm going to let you test your containers now. Test their capacities by carefully spooning or pouring the rice from one to another. See if you can find the container on your tray that has the biggest capacity and the container with the smallest capacity. Draw them on your recording sheet. (Show students relevant sections on the sheet.) If you spill, just scoop the rice off your tray and put it back. (Allow ample time for experimentation.)

T: Hold up the container on your tray that has the biggest capacity. (Observe whether or not the students exhibit understanding.) How did you know? (Discuss reasonable answers.) Hold up the container with the smallest capacity. How could you tell? (Check for understanding; discuss reasonable answers.)

> **A NOTE ON MULTIPLE MEANS OF ENGAGEMENT:**
>
> Scaffold the lesson for students who struggle by giving them extra practice with capacity using interactive technology such as the game found at http://education.abc.net.au/home#!/games/-/mathematics.

Problem Set (10 minutes)

In this lesson, the *capacity recording sheet* for the activity serves as the primary Problem Set for the Concept Development. There is an optional Problem Set, which can be used to ignite discussion by having students consider the capacities of pairs of objects and then wonder if the containers such as the teapot are necessarily completely filled.

Lesson 13: Compare volume using *more than*, *less than*, and *the same as* by pouring.

Student Debrief (6 minutes)

Lesson Objective: Compare volume using *more than*, *less than*, and *the same as* by pouring.

The Student Debrief is intended to invite reflection and active processing of the total lesson experience.

Invite students to review their solutions for the Recording Sheet. They should check work by comparing answers with a partner before going over answers as a class. Look for misconceptions or misunderstandings that can be addressed in the Student Debrief. Guide students in a conversation to debrief the Recording Sheet and process the lesson.

Any combination of the questions below may be used to lead the discussion.

- How were we comparing today? Were we comparing length, weight, or the number of objects?
- What does the word **capacity** mean to you?
- Which of your containers had the biggest capacity?
- Which had the smallest capacity?
- Did the shape of the container make a difference in how much it could hold?
- Were you surprised by anything you learned during this activity?

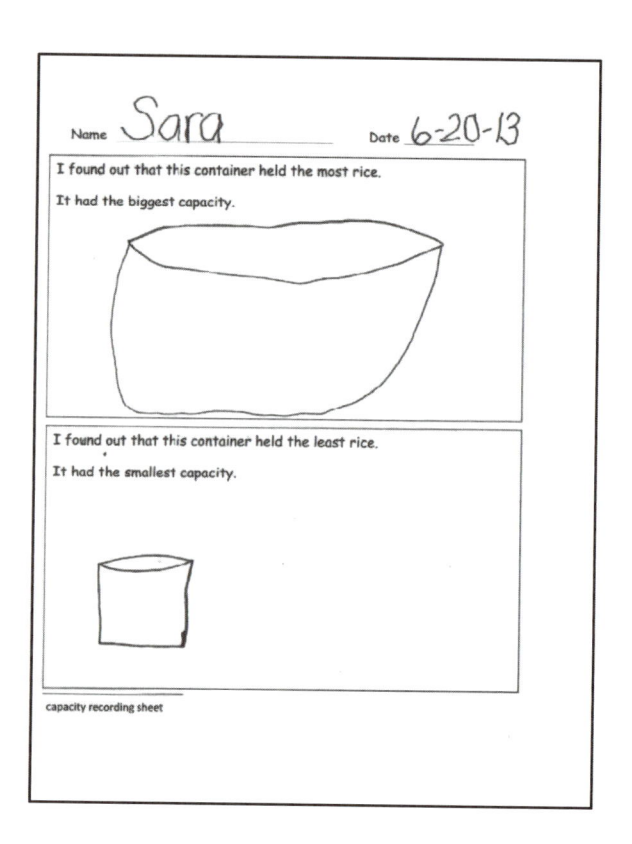

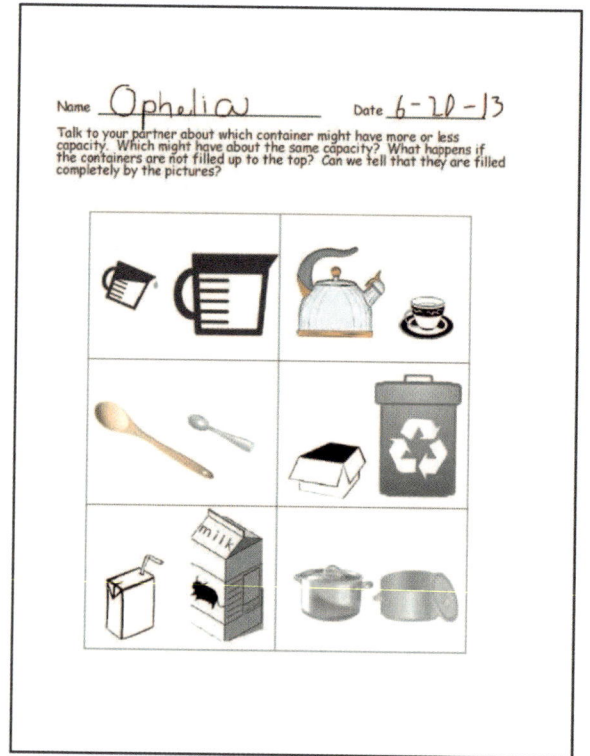

A STORY OF UNITS   Lesson 13 Problem Set   K•3

Name _____   Date _____

Talk to your partner about which container might have more or less capacity. Which might have about the same capacity? What happens if the containers are not filled up to the top? Can we tell that they are filled completely from looking at the pictures?

Lesson 13: Compare volume using *more than*, *less than*, and *the same as* by pouring.

Name _____ Date _____

In class, we have been working on capacity. Encourage your child to explore with different-sized containers to see which ones have the biggest and smallest capacity. Children can experiment by pouring liquid from one container to another.

All the homework you will see for the next few days will be a review of skills taught from Module 1.

Each rectangle shows 6 objects. Circle 2 different sets within each. The two sets represent the two parts that make up the 6 objects. The first one has been done for you.

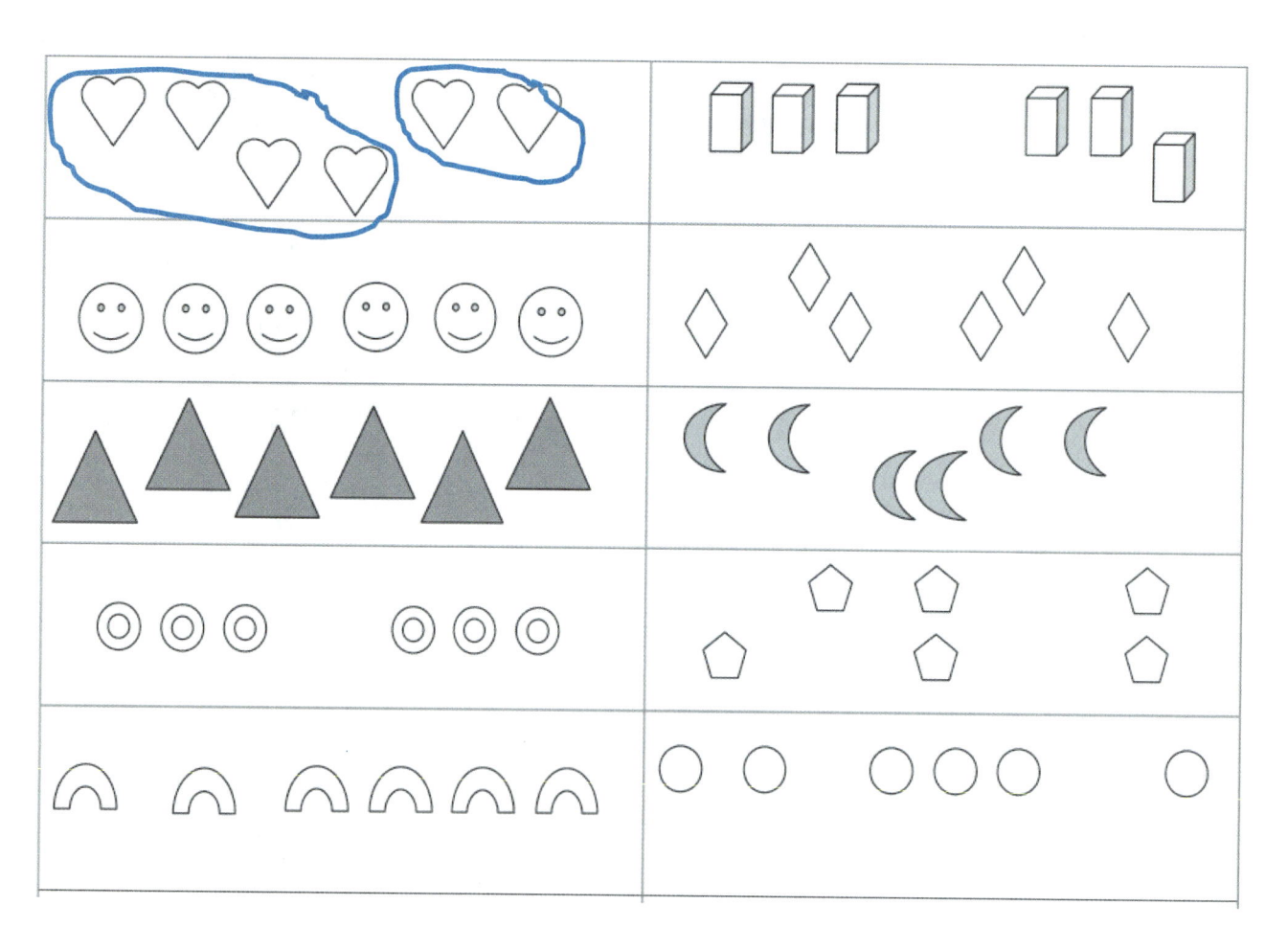

Name _____   Date _____

I found out that this container held the most rice.

It had the biggest capacity.

I found out that this container held the least rice.

It had the smallest capacity.

capacity recording sheet

A STORY OF UNITS — Lesson 13 Fluency Template — K•3

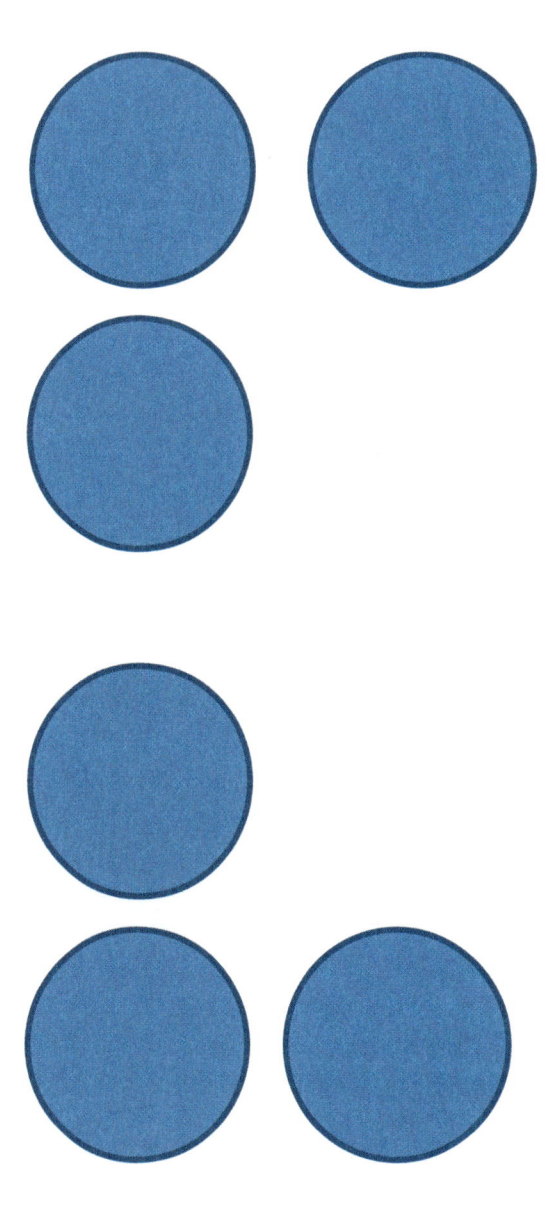

dot cards of 6

Lesson 13: Compare volume using *more than*, *less than*, and *the same as* by pouring.

dot cards of 6

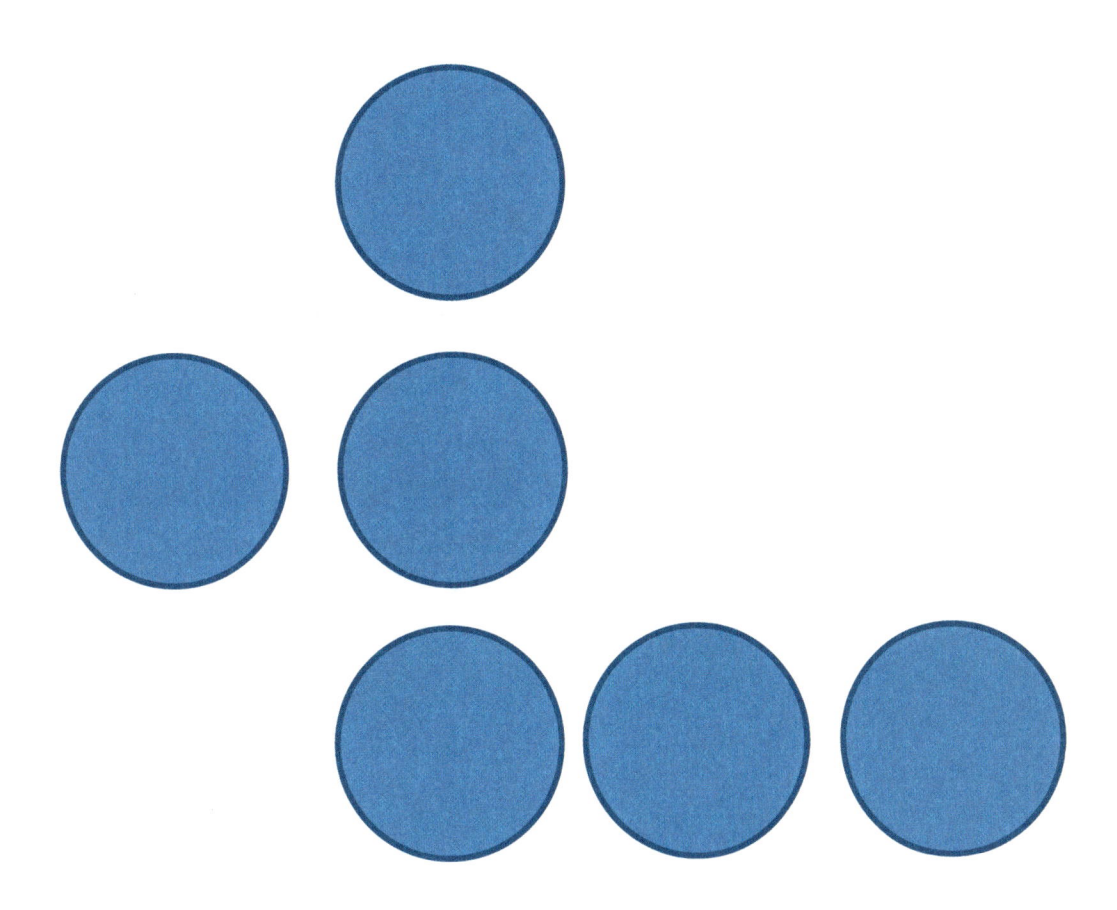

dot cards of 6

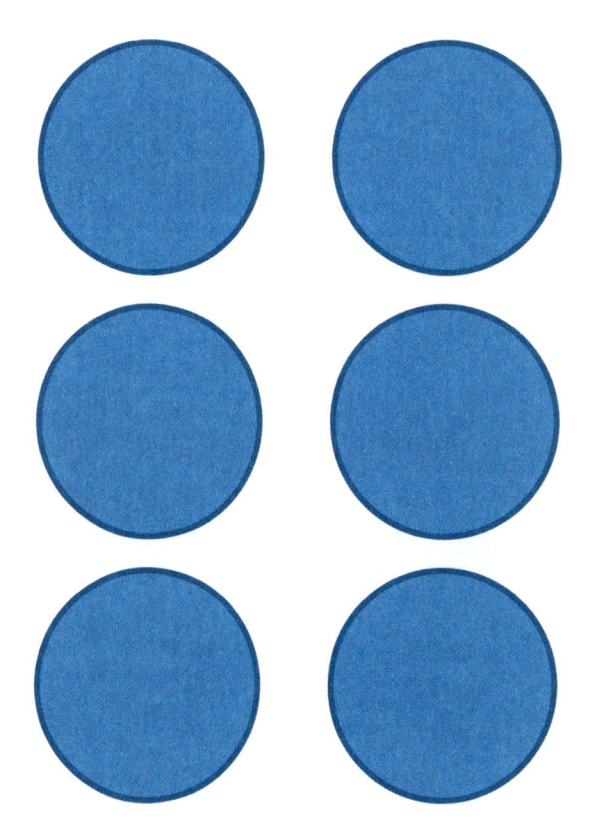

Lesson 13: Compare volume using *more than*, *less than*, and *the same as* by pouring.

dot cards of 6

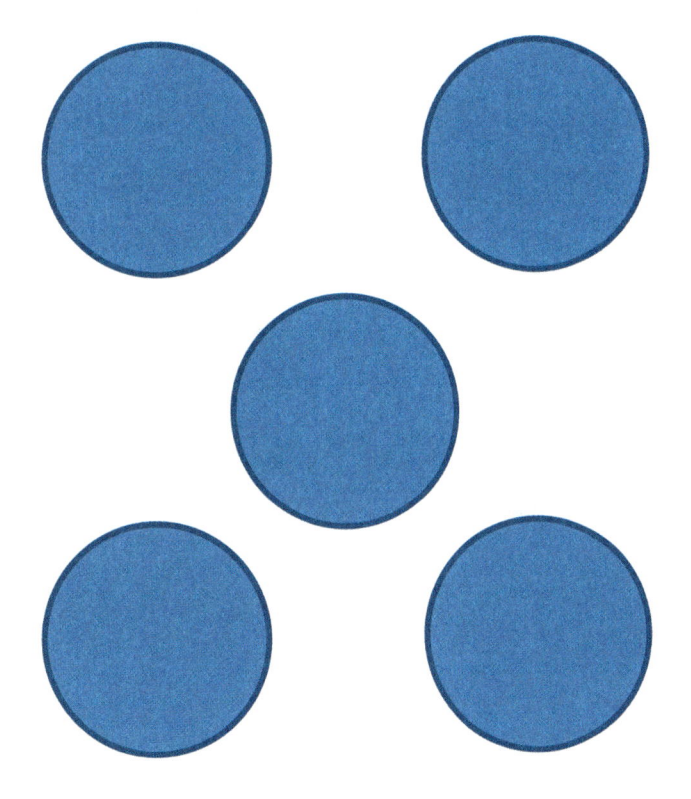

Lesson 13: Compare volume using *more than*, *less than*, and *the same as* by pouring.

dot cards of 6

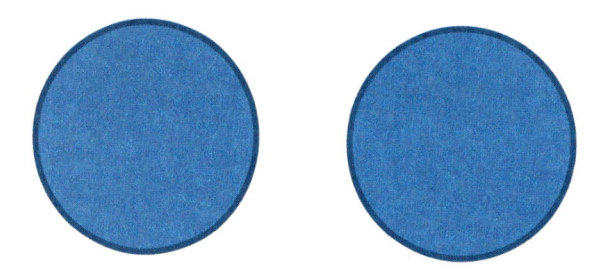

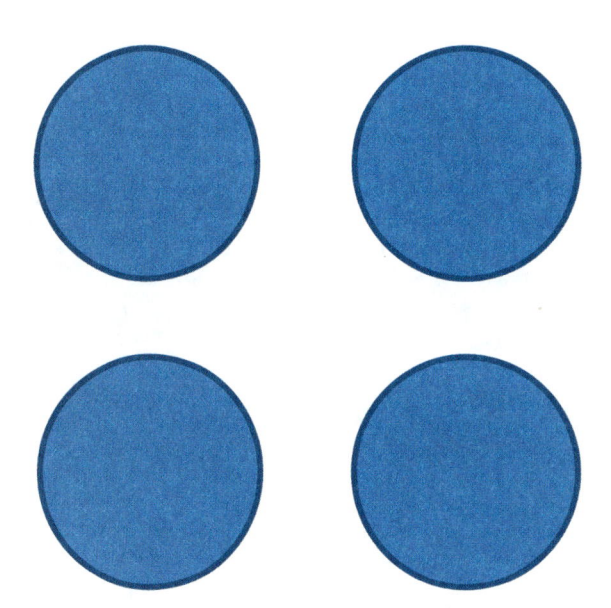

dot cards of 6

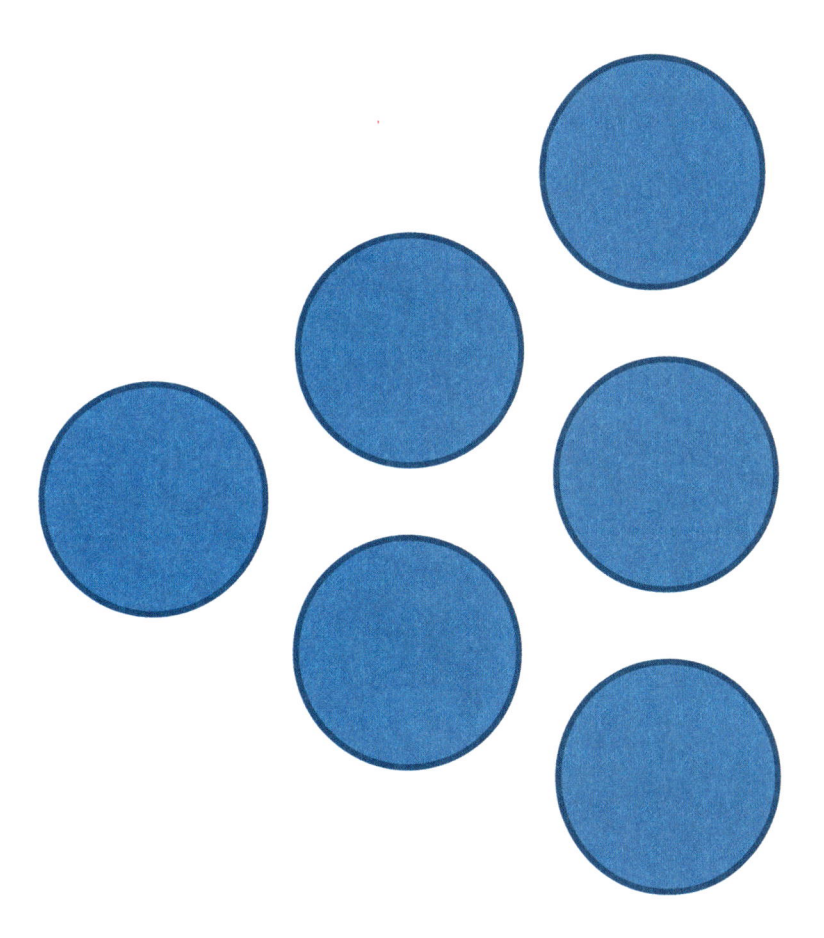

dot cards of 6

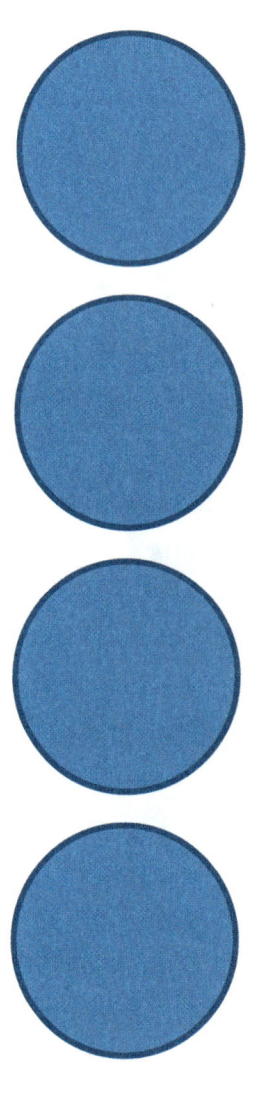

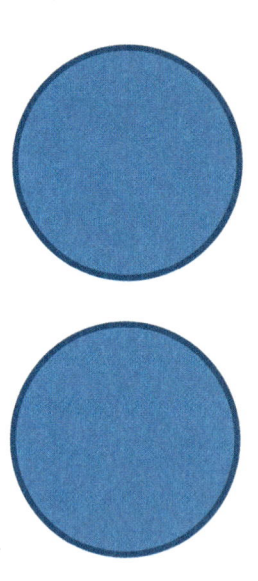

A STORY OF UNITS Lesson 13 Fluency Template K•3

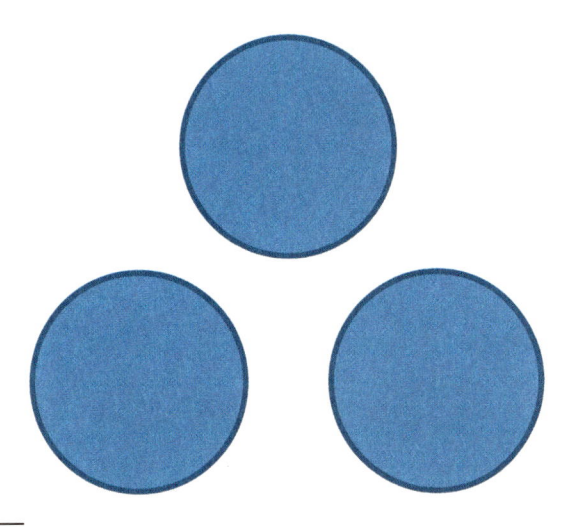

---

dot cards of 6

Lesson 13: Compare volume using *more than*, *less than*, and *the same as* by pouring.

127

dot cards of 6

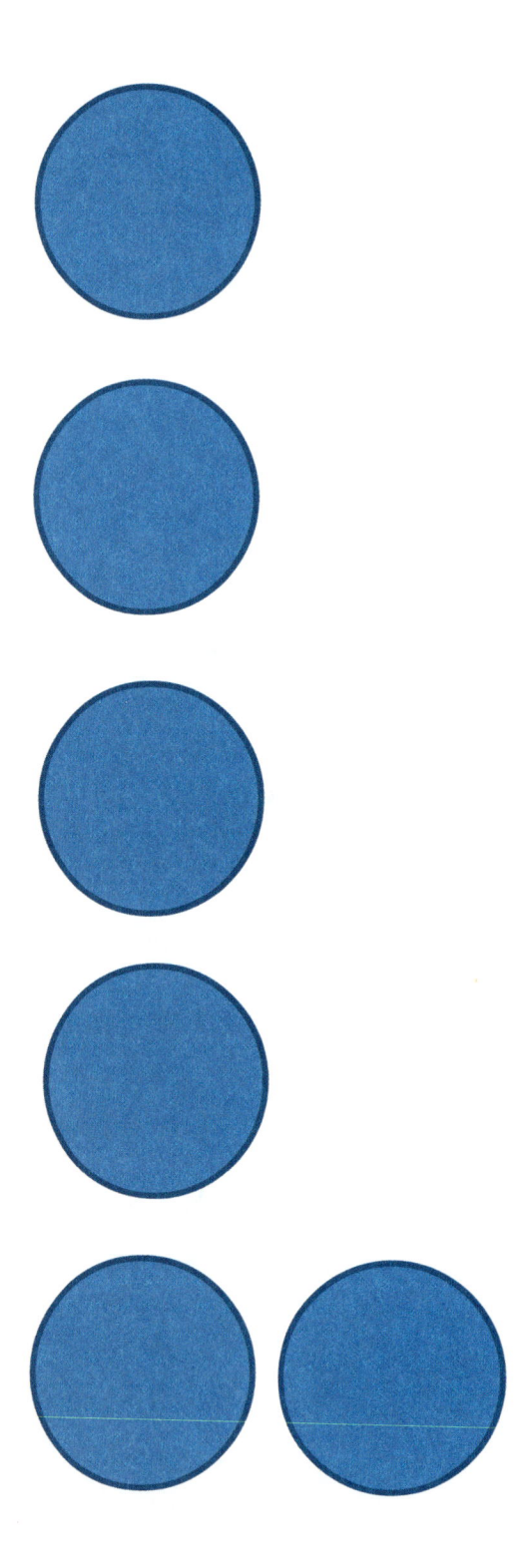

| A STORY OF UNITS | Lesson 14 | K•3 |

Lesson 14

Objective: Explore conservation of volume by pouring.

Suggested Lesson Structure

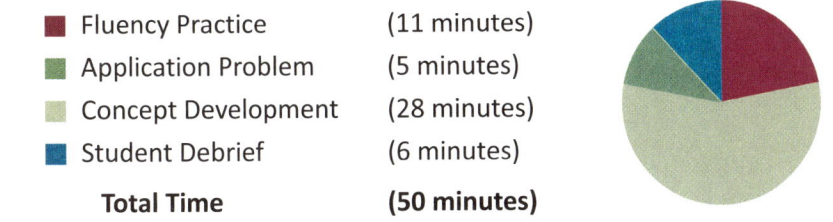

- **Fluency Practice**   (11 minutes)
- **Application Problem**   (5 minutes)
- **Concept Development**   (28 minutes)
- **Student Debrief**   (6 minutes)
- **Total Time**   **(50 minutes)**

Fluency Practice (11 minutes)

- Say Ten Push-Ups **K.NBT.1**   (3 minutes)
- Hidden Numbers (10 as the Whole) **K.OA.3**   (5 minutes)
- Double 5-Groups **K.CC.2**   (3 minutes)

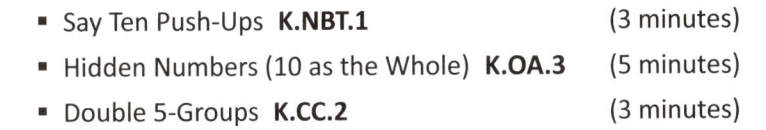

Say Ten Push-Ups (3 minutes)

Conduct the activity as outlined in Lesson 1. Continue to 20 (2 ten, or 10 and 10).

Hidden Numbers (10 as the Whole) (5 minutes)

Materials: (S) Hidden numbers mat (Lesson 3 Fluency Template)

Conduct the activity as described in Lesson 3, except students do not need to cross out any of the fish. Guide them to find twos, threes, fours, and fives within the larger group of 10.

Double 5-Groups (3 minutes)

Conduct the activity as described in Lesson 10, but now continue to 20 (2 ten, or 10 and 10).

Application Problem (5 minutes)

Materials: (S) Small ball of clay

With your clay, make a bowl big enough to hold a yummy strawberry. Now, make a little vase just the right size for a tiny flower. Which one do you think would have more capacity?

Compare your containers to those of your friend's. Do they look alike? Do you think hers would have more capacity?

Note: This Application Problem leads students to think about the effect of the shape of a container on its volume or the appearance thereof, serving as an anticipatory set for today's lesson. Circulate during the exercise to encourage correct use of vocabulary.

Lesson 14:   Explore conservation of volume by pouring.

A STORY OF UNITS

Lesson 14 K•3

Concept Development (28 minutes)

Materials: (T) Set of student materials for demonstration (S) 2 cups of rice, clear containers (if possible) with varying diameters (e.g., a glass, small bowl, small vase with an interesting shape, bottle, mug), tray, funnel, spoon, volume recording sheet (Template)

Note: Save a set of student materials for the culminating task in Lesson 32.

MP.7

T: In the last lesson, we talked about the capacities of our containers. I wonder what the capacity of this bowl is. How could I find out?

S: You could fill it with rice.

T: Tell me when to stop! (Use a spoon to fill the bowl.) There. Let me draw how the rice looks in this bowl on my recording sheet. (Demonstrate.)

T: Look at this bottle. I wonder if the capacity of the bottle is more or less than the capacity of the bowl. How could we find out?

S: Pour the rice into the bottle!

T: Good idea! I will use this funnel so I don't lose any. (Pour the rice into the bottle.) What do you notice?

S: The bottle isn't full! The rice only goes partway up the side!

T: Hmmm. I didn't spill any. What do you think happened?

S: The bottle is taller, so the rice doesn't look as big. → It must hold more.

T: Yes. The capacity of the bottle is more than the capacity of the bowl. Let me draw how the rice looks in the bottle. (Draw.) What will happen if I pour the rice back into the bowl?

S: It will be full again!

> **NOTES ON MULTIPLE MEANS OF ACTION AND EXPRESSION:**
>
> Encourage students working above grade level to explain their reasoning about capacity. Ask them to explain why they think one container holds more or less rice than another. Have them explain what happens to the rice as it moves from one container to the next. Ask them to write down their reasoning and share it with a friend.

T: Let's test your guess. (Pour the rice back into the bowl.) You were right! I'm going to let you experiment with your containers now. Fill your small bowl with the rice, and then notice how that same amount of rice looks in the other containers. On your recording sheet, draw what you see. Pour the rice carefully so that you don't lose any between containers. If you do, scoop it up from your tray, and put it in to make sure that your tests are fair! (Allow ample time for experimentation and discussion.)

T: Who would like to share something they learned during the experiment?

S: The rice looked really tall in this one! → This one looks like it could hold a lot more. It almost looked empty!

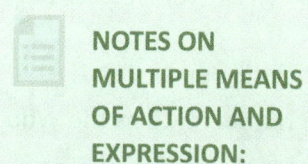

Lesson 14: Explore conservation of volume by pouring.

A STORY OF UNITS
Lesson 14 K•3

Problem Set (10 minutes)

In this lesson, the volume recording sheet serves as the Problem Set for the Concept Development.

Student Debrief (6 minutes)

Lesson Objective: Explore conservation of volume by pouring.

The Student Debrief is intended to invite reflection and active processing of the total lesson experience.

Invite students to review their recording sheets. They should check work by comparing answers with a partner before going over answers as a class. Look for misconceptions or misunderstandings that can be addressed in the Student Debrief. Guide students in a conversation to process the lesson.

Any combination of the questions below may be used to lead the discussion.

- Look at your Recording Sheet. In which container did it look like you had the most rice?
- In which container did it look like you had the least rice?
- Did the amount of the rice ever change?
- Were the shapes of the containers the same? Describe them to your partner.
- Does the shape of the container make the amount of the rice seem different? Why?
- What math vocabulary did we use today to communicate precisely?

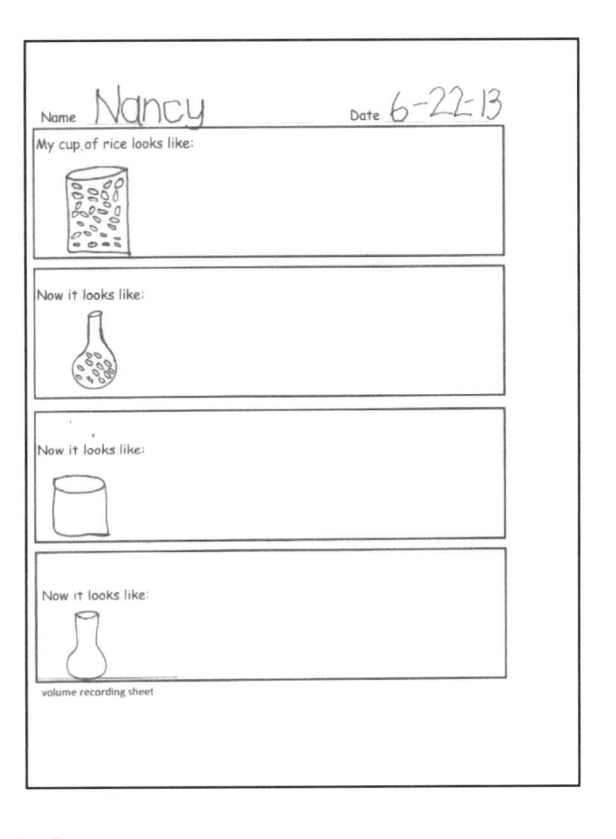

Lesson 14: Explore conservation of volume by pouring.

EUREKA MATH

131

Name _____ Date _____

Within each rectangle, make one set of 6 objects. The first one has been done for you.

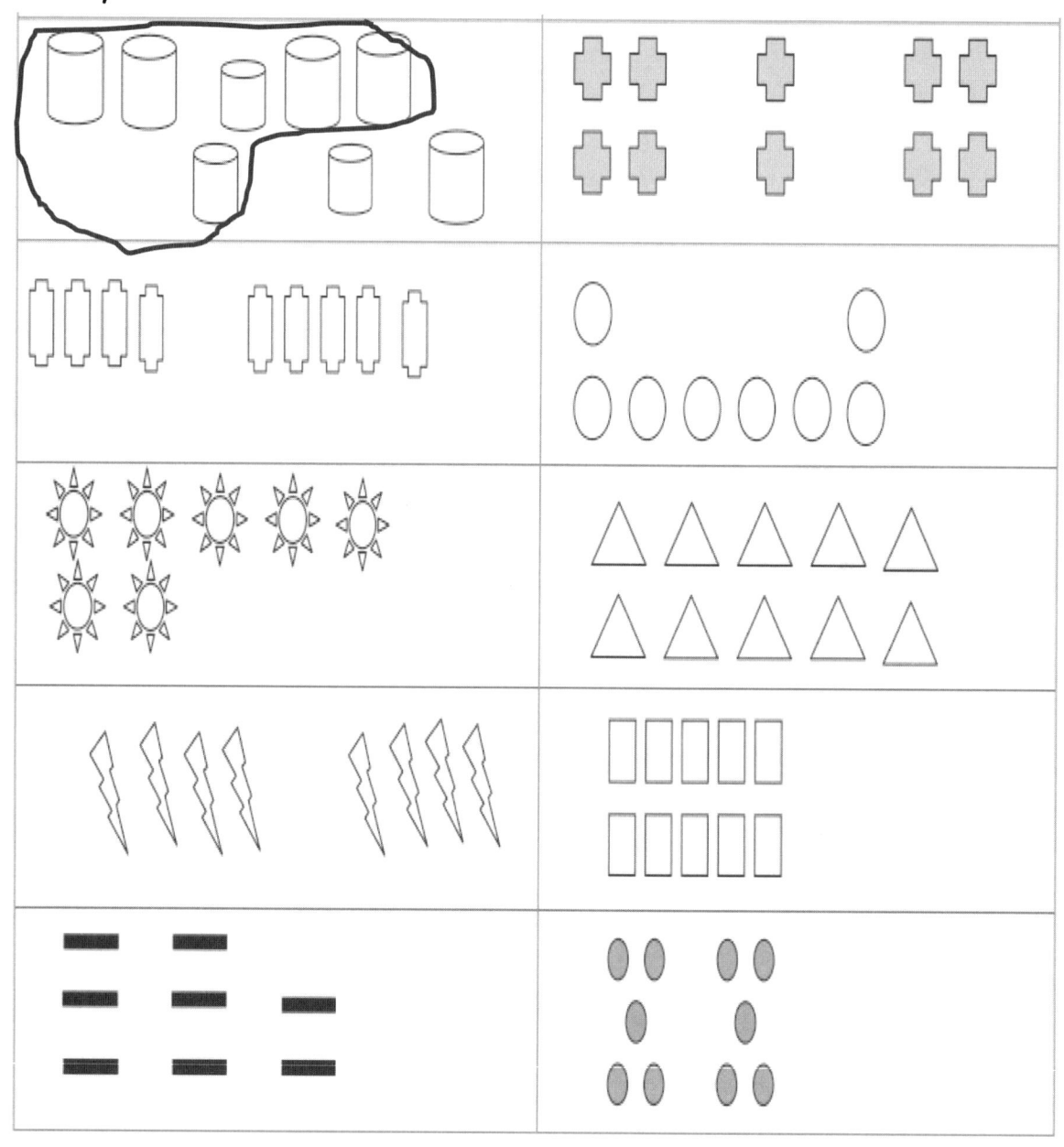

A STORY OF UNITS   **Lesson 14 Template**   K•3

Name _____ Date _____

My cup of rice looks like:

Now it looks like:

Now it looks like:

Now it looks like:

volume recording sheet

Lesson 14: Explore conservation of volume by pouring.

Lesson 15

Objective: Compare using *the same as* with units.

Suggested Lesson Structure

- Fluency Practice   (12 minutes)
- Application Problem   (5 minutes)
- Concept Development   (27 minutes)
- Student Debrief   (6 minutes)

Total Time   **(50 minutes)**

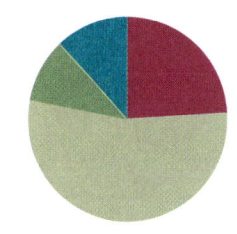

Fluency Practice (12 minutes)

- Dot Cards of 7 **K.CC.5, K.CC.2**   (4 minutes)
- Make It Equal **K.CC.6**   (3 minutes)
- Building *1 More* and *1 Less* Towers **K.CC.4c** (5 minutes)

Dot Cards of 7 (4 minutes)

Materials: (T/S) Dot cards of 7 (Fluency Template)

Note: This activity deepens students' knowledge of embedded numbers and develops part–whole thinking, crucial to the work of the upcoming modules.

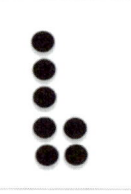

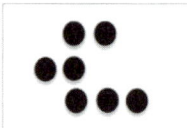

 T: (Show 7 dots.) How many do you see? (Give students time to count.)
 S: 7.
 T: How can you see 7 in two parts?
 S: (Point to the card.) 5 here and 2 here. → I see 3 here and 4 here.

Continue with other cards of 7. Distribute the cards to the students for partner sharing time. Have them pass the cards at a signal.

Make It Equal (3 minutes)

Materials: (S) Bag of beans, foam or laminated paper work mat, 2 dice with 6-dot side covered

Note: In this activity, students experience comparison visually, a skill foundational to the work of this module.

1. Teacher introduces the term *equal* as meaning *the same number*.
2. Both partners roll dice and put that many beans on their mat.
3. Partner A makes her beans equal to her partner's by taking off or putting on more beans.
4. Partner B counts to verify.
5. Switch roles and play again.

Building *1 More* and *1 Less* Towers (5 minutes)

Materials: (S) 10 linking cubes

Note: In this activity, students connect increasing and decreasing height to increasing and decreasing numerical value.

Conduct the activity as described in Lesson 13, but now challenge students to stop at a certain number and then change directions so that they state the pattern of *1 more* or *1 less* starting from numbers other than 1 or 10.

T: Build up your tower while saying "1 more." Stop when you get to 5.
S: 1. 1 more is 2. 2. 1 more is 3. 3. 1 more is 4. 4. 1 more is 5.
T: Stop! Now, take it apart while saying 1 less. Stop when you get to 3.
S: 5. 1 less is 4. 4. 1 less is 3.
T: Stop!

Continue changing directions several more times. It might be helpful to use a stick of cubes that show a color change at 5 to facilitate identifying the number of cubes in the tower.

Application Problem (5 minutes)

Materials: (S) Small ball of clay and 10 beans

Use your clay to make a container just large enough to hold your 10 beans. Test to see if the beans fit! Show your work to your partner.

Note: In this exercise, students' thinking expands to consider that volume can be measured in units, in this case by beans. This serves as an anticipatory set for today's introduction to comparing volume through units.

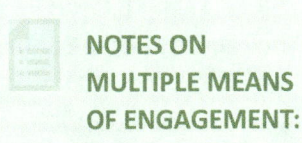

NOTES ON MULTIPLE MEANS OF ENGAGEMENT:

Extend the thinking of those students working above grade level by asking them to think about what else would fit in their container. Allow them to experiment with materials in the classroom. Ask them to estimate how much of their chosen material would fit into their container, and ask them to explain why they were correct or incorrect.

Lesson 15: Compare using the same as with units.

A STORY OF UNITS　　　　　　　　　　　　　　　　　　　　　　　　　　　　　Lesson 15　K•3

Concept Development (27 minutes)

Materials: (T) Set of student materials for demonstration, we've got the scoop recording sheet (Template) affixed to white board (S) 2 cups of rice, assortment of containers (teacup, small bottle, bowl, glass, small box, measuring cup), small scoop such as a coffee scoop, funnel, and tray per pair or small group; we've got the scoop recording sheet (Template)

T: (Hold up the scoop.) I wonder how many of these little scoops of rice it will take to fill my teacup. Does anyone have a guess?

S: (Various responses.)

T: I will put in 1 scoop so you can see how it looks in the cup. Watch how I am careful to level off the scoop before I pour it. It's not fair using scoops that are only half full! (Model correct measuring technique.) Do you want to change your guess?

S: (Various responses.)

T: Student A, would you please help me finish filling my cup? Let's count with Student A while he uses the scoop to fill the teacup.

S: 2, 3, 4, 5, …, 10. Ten scoops!

T: It took 10 scoops to fill the teacup. 10 scoops is the same as 1 teacup of rice! Let me put that on my recording sheet. (Demonstrate.)

T: (Hold up a smaller container.) How many scoops do you think it will take to fill this? Will it still be 10?

S: That one is smaller. → It will take 5 scoops. → I think it will take 7.

T: Student B, would you please come up to help? Count with Student B as he uses the scoop to fill the container.

S: 1, 2, 3, 4, 5, 6. It took 6 scoops. It holds less!

T: This container holds the same amount as 6 scoops. The capacity of this container is the same as 6 scoops. I will record that on my sheet, too. (Demonstrate.)

T: I want you to work with your partner to find out how many scoops each of the containers on your tray holds. Count the scoops, and fill each container carefully. Use your funnel if you need to. Each time, remember to fill the scoop up all the way, but make sure it isn't spilling over. Write your discoveries on your recording sheet. (Allow time for measurement and experimentation.)

T: Put your things back on your tray. Who would like to share something on his or her recording sheet? (Allow time for discussion.)

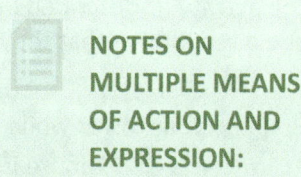

MP.6

NOTES ON MULTIPLE MEANS OF ACTION AND EXPRESSION:

Scaffold the lesson for English language learners by using motions. For example, hold up the scoop when directing students to count the scoops it takes to fill their containers, and hold up the funnel when directing students to use the funnel if they need it.

A STORY OF UNITS • Lesson 15 • K•3

Problem Set (10 minutes)

In this lesson, the we've got the scoop recording sheet serves as the Problem Set for the Concept Development.

Student Debrief (6 minutes)

Lesson Objective: Compare using *the same as* with units.

The Student Debrief is intended to invite reflection and active processing of the total lesson experience.

Invite students to review their recording sheets. Students should check work by comparing answers with a partner before going over answers as a class. Look for misconceptions or misunderstandings that can be addressed in the Debrief. Guide students in a conversation to process the lesson.

Any combination of the questions below may be used to lead the discussion.

- Which of your containers held the most scoops?
- Which of your containers held the least number of scoops?
- Which container had the largest capacity? How did you know?
- Which container do you think had the least capacity? How did you know?
- Do you notice any patterns from your work today?
- Did you make any surprising discoveries during your work today?

Lesson 15: Compare using the same as with units.

137

Name _____   Date _____

Circle 2 sets within each set of 7. The first one has been done for you.

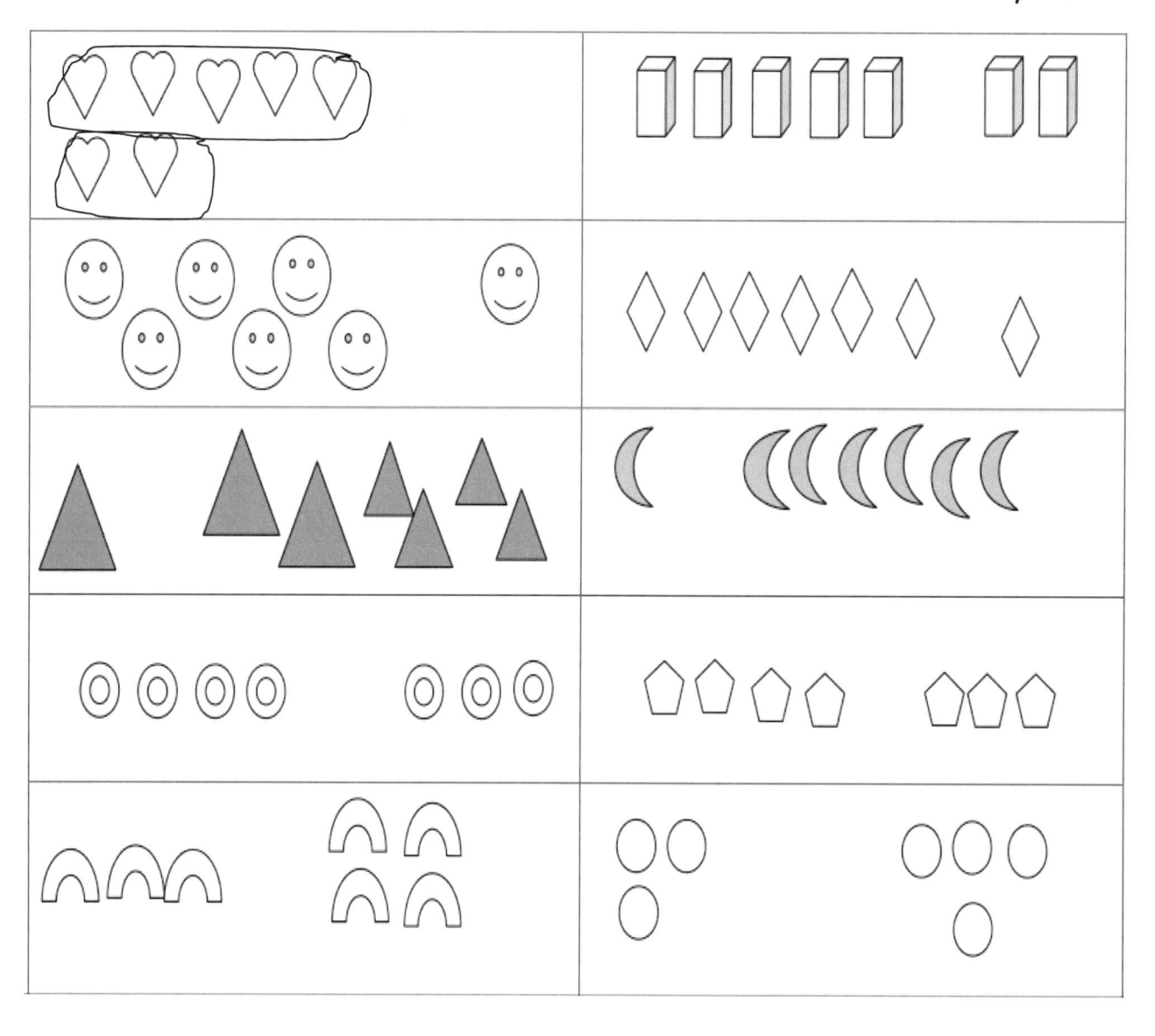

Lesson 15: Compare using the same as with units.

A STORY OF UNITS

Lesson 15 Template   K•3

Name _____   Date _____

We've Got the Scoop!

_____ is the same as _____ scoops.

_____ is the same as _____ scoops.

_____ is the same as _____ scoops.

_____ scoops is the same as _____

_____ scoops is the same as _____

we've got the scoop recording sheet

Lesson 15:   Compare using the same as with units.

dot cards of 7

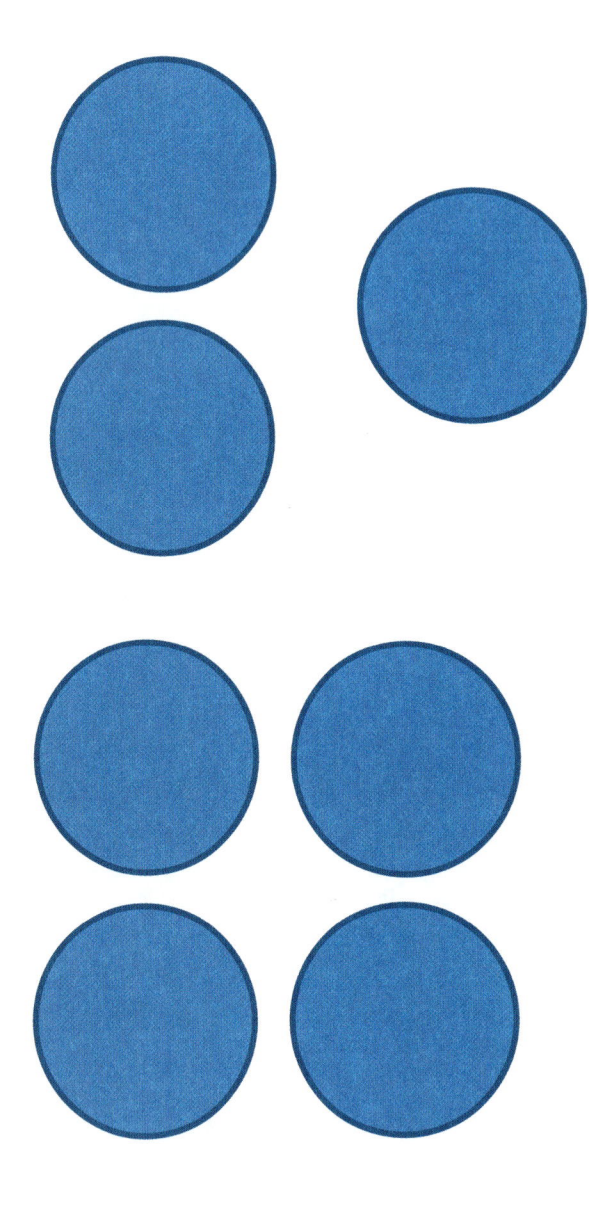

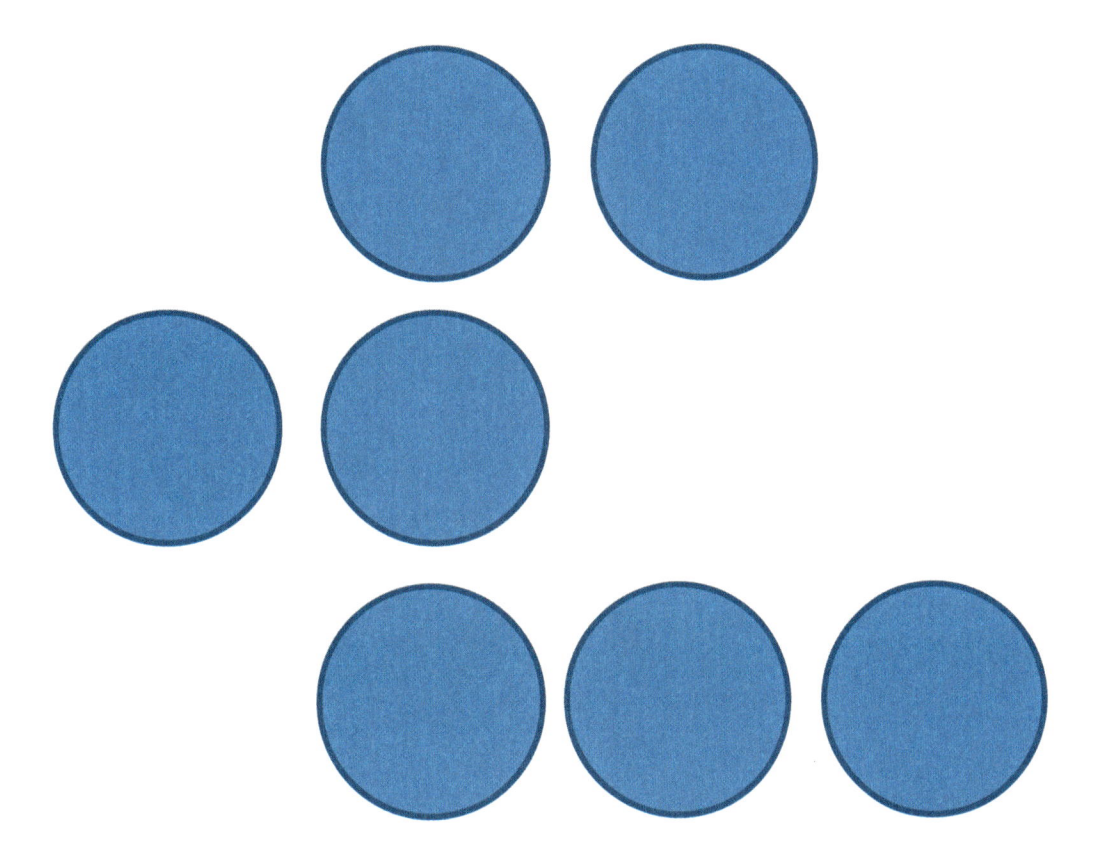

A STORY OF UNITS — Lesson 15 Fluency Template — K•3

dot cards of 7

Lesson 15: Compare using the same as with units.

141

dot cards of 7

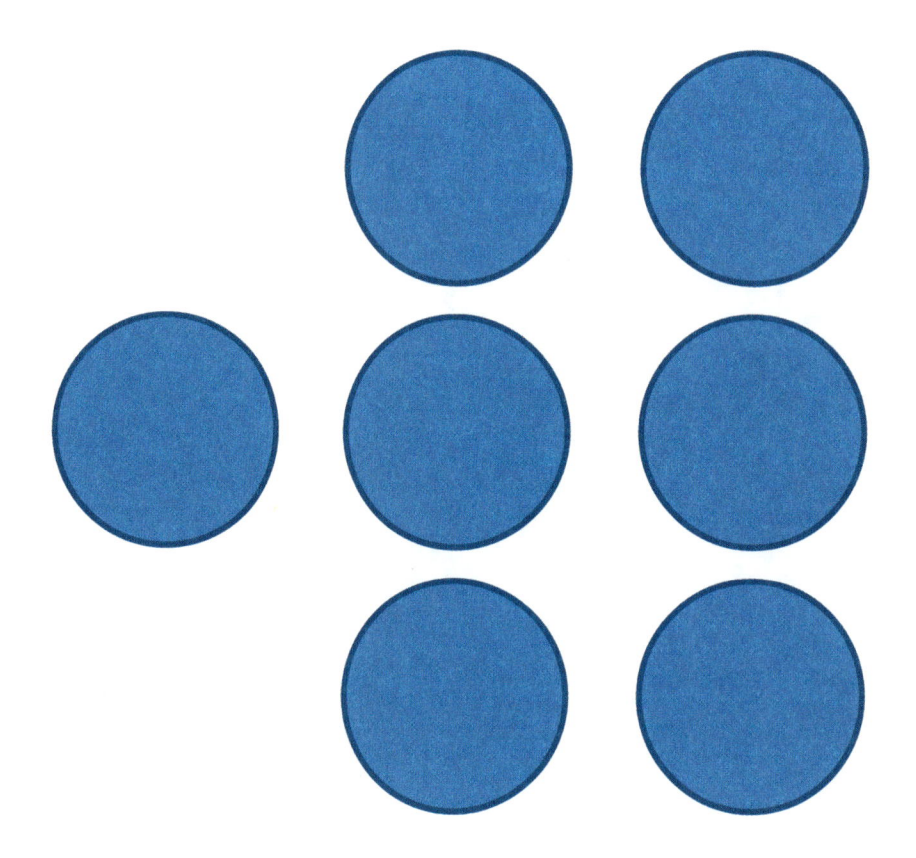

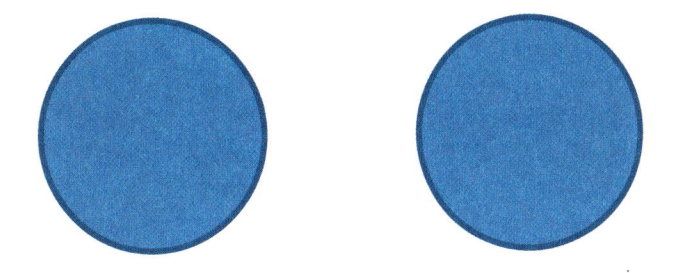

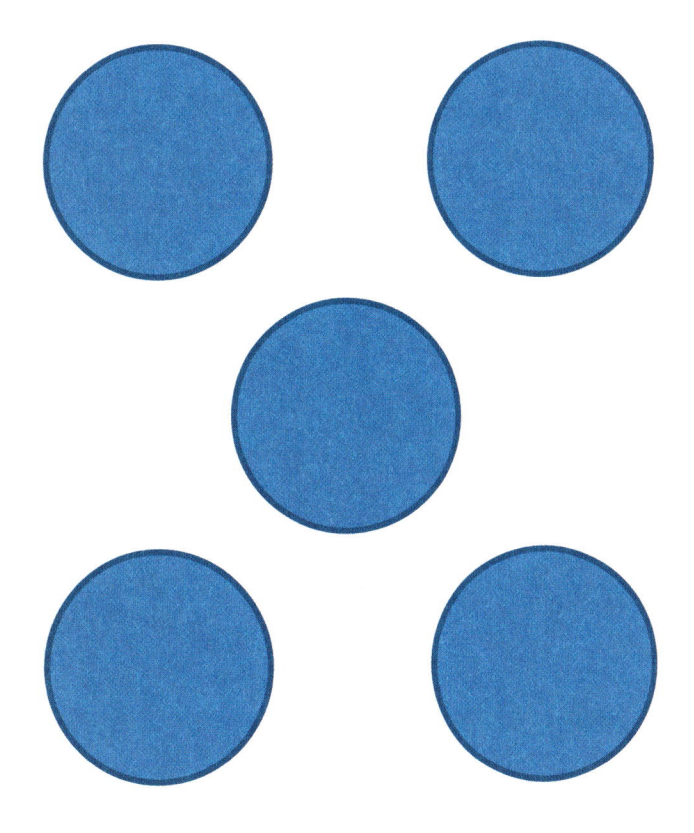

dot cards of 7

Lesson 15: Compare using the same as with units.

A STORY OF UNITS

Lesson 15 Fluency Template K•3

dot cards of 7

144   Lesson 15:   Compare using the same as with units.

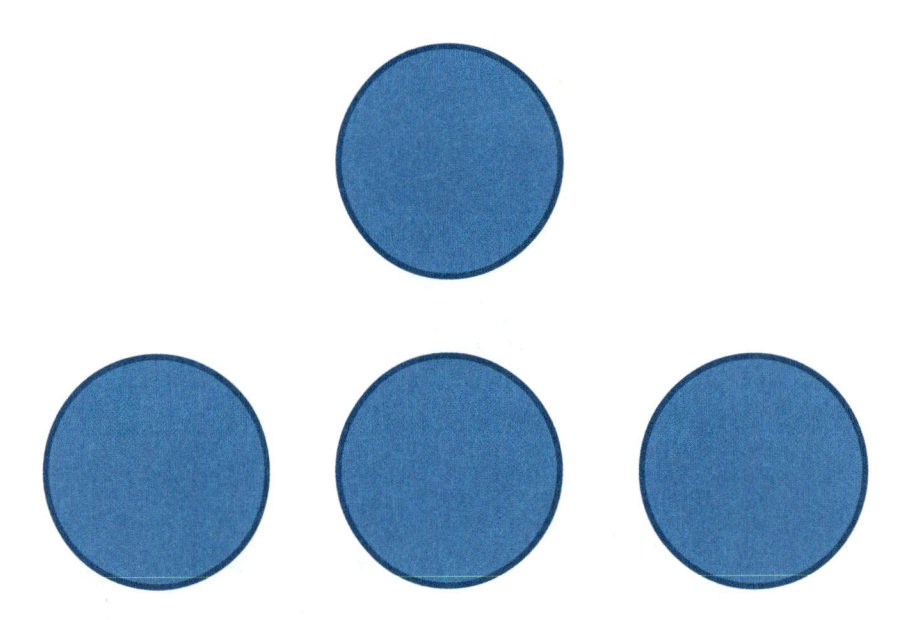

© 2015 Great Minds. eureka-math.org
GK-M3-TE-B3-1.3.1-01.2016

EUREKA MATH

dot cards of 7

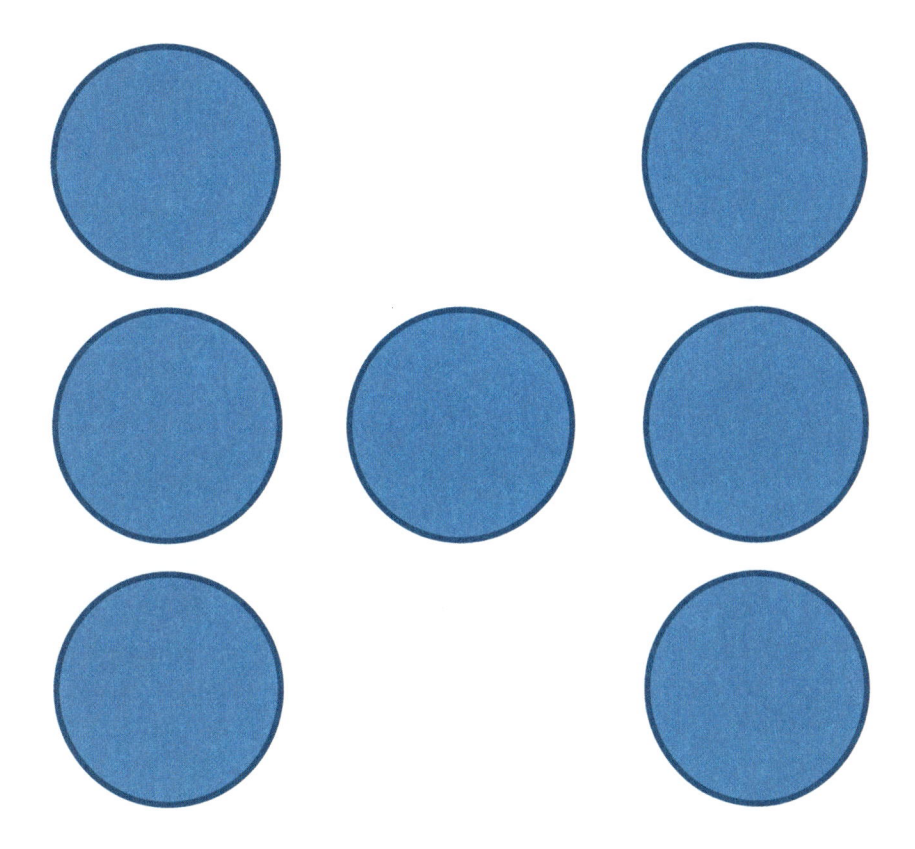

dot cards of 7

Lesson 15: Compare using the same as with units.

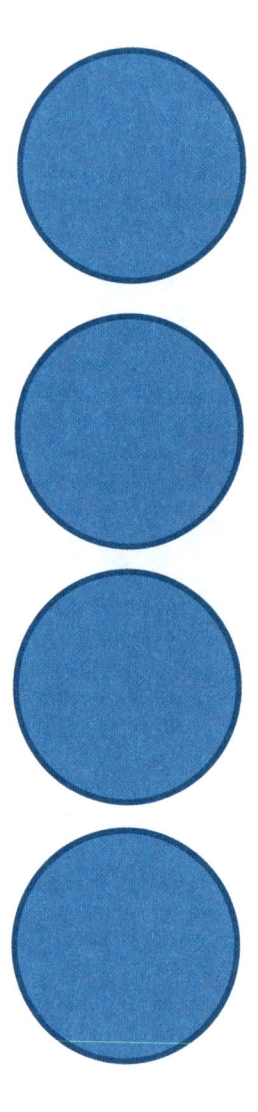

dot cards of 7

Lesson 15: Compare using the same as with units.

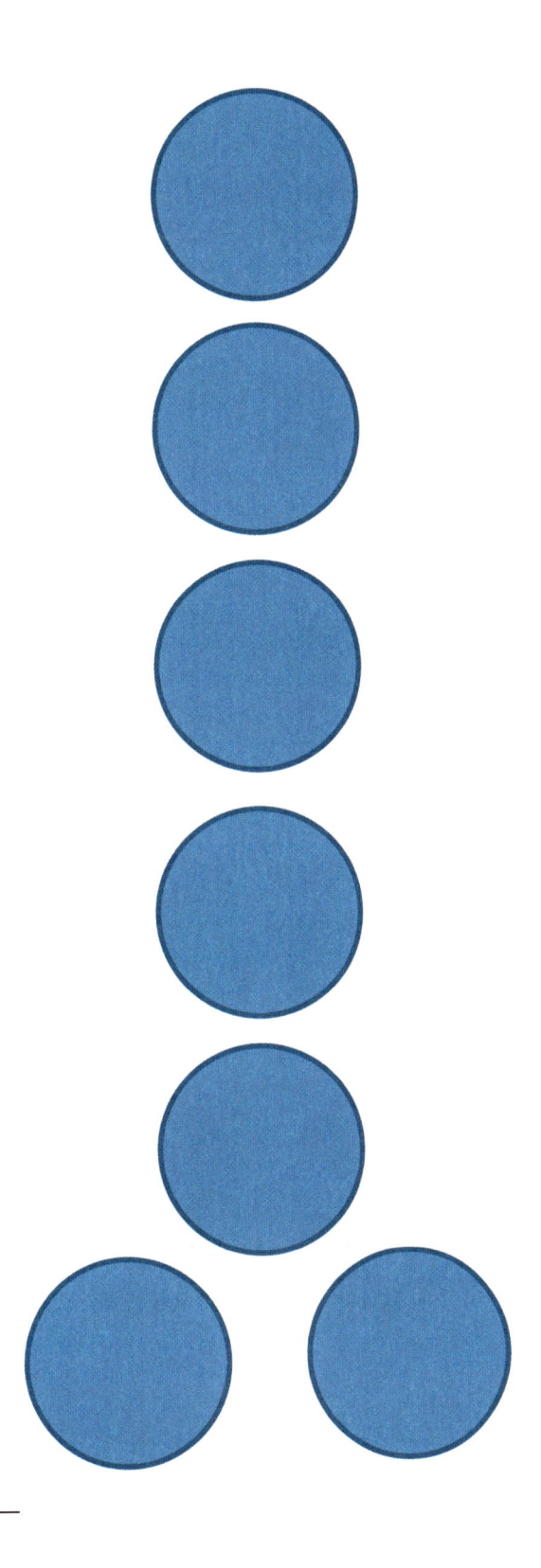

dot cards of 7

Lesson 15: Compare using the same as with units.

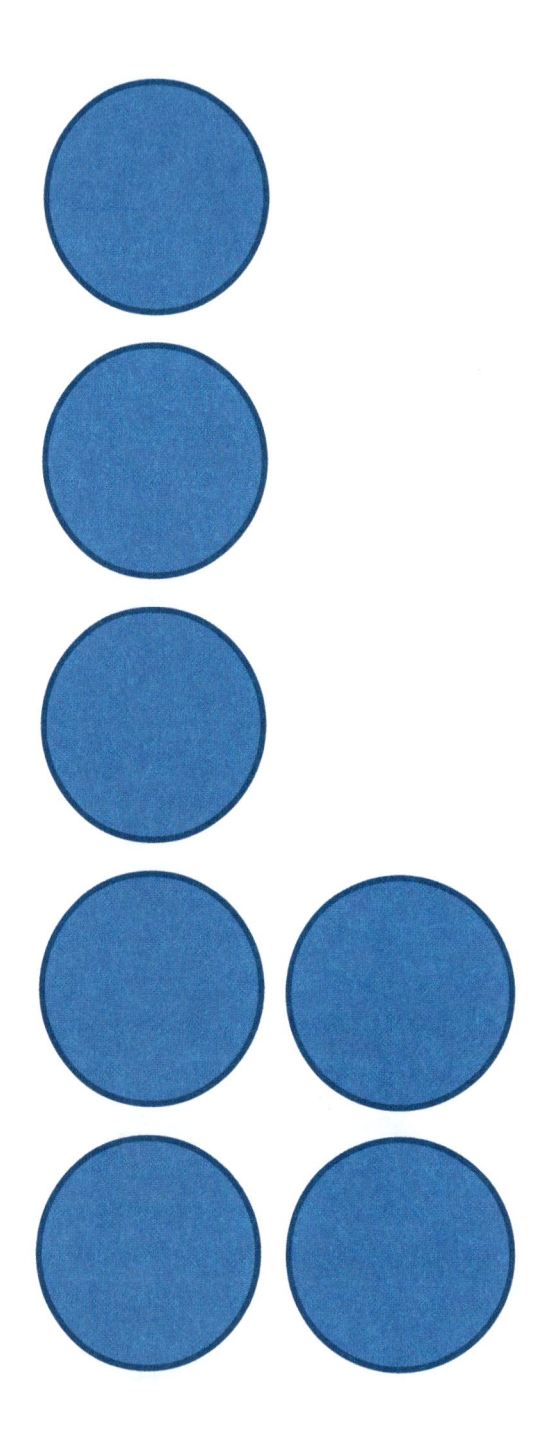

Kindergarten Mid-Module 3 Assessment (Administer after Topic D)
Kindergarten End-of-Module 3 Assessment (Administer after Topic H)

Assessment time is an important component of the student–teacher relationship. In early grades, it is especially important to establish a positive and collaborative attitude when analyzing progress. Sit next to the student rather than opposite, and support the student in understanding the benefits of sharing and examining her level of mastery.

Please use the specific language of the assessment and, when possible, translate for English language learners. (This is a math rather than a language assessment.) If a student is unresponsive, wait about 15 seconds for a response. Record the student's results in two ways: (1) the narrative documentation after each topic set and (2) the overall score per topic using *A Progression Toward Mastery*. Use a stopwatch to document the elapsed time for each response.

Within each assessment, there is a set of problems targeting each topic. Each set comprises three or four related questions. Document what the student did and said in the narrative, and use the rubric for the overall score for each set.

If the student is unable to perform any part of the set, his score cannot exceed Step 3. However, if the student is unable to use his words to tell what he did, do not count that against the student quantitatively. Be aware that an English language learner's ability to articulate compared to other students will likely be significantly different. If the student asks for or needs a hint or significant support, provide it, but the score is automatically lowered. This ensures that the assessment provides a true picture of what a student can do independently.

If a student scores at Step 1 or 2, repeat that topic set again at two-week intervals, noting the date of the reassessment in the space at the top of the student's record sheet. Document progress on this one form. If the student is very delayed in her response but completes it, reassess to determine if there is a change in the time elapsed.

House the assessments in a three-ring binder or student portfolio. By the end of the year, there will be 10 assessments for each student. Modules 1, 3, 4, and 5 have two assessments each, whereas Modules 2 and 6 only have one. Use the *Class Record Sheet* (following the rubric) for an easy reference to assess students' strengths and weaknesses.

These assessments can be valuable for daily planning, parent conferences, and Grade 1 teacher preparation to receive these students.

A STORY OF UNITS

Mid-Module Assessment Task K•3

Student Name: _____

Topic A: Comparison of Length and Height

Rubric Score: _____ Time Elapsed: _____

	Date 1	Date 2	Date 3
Topic A			
Topic B			
Topic C			
Topic D			

Materials: (S) 6- and 9-inch pieces of string

Cover strings so each string has 3 inches exposed from a piece of paper. Let pieces be parallel to each other.

1. Each piece of string is hiding under the paper. Can we tell which one is longer? Why or why not?
2. (Uncover them.) Compare this string to this string. Use the words *longer than*.
3. Move the strings so that they line up on one end.
4. Compare these strings now. Use the words *shorter than*.
5. When we use the words *longer than* or *shorter than*, what are we comparing?

What did the student do?	What did the student say?
1.	
2.	
3.	
4.	
5.	

Module 3: Comparison of Length, Weight, Capacity, and Numbers to 10

A STORY OF UNITS

Mid-Module Assessment Task K•3

Topic B: Comparison of Length and Height of Linking Cube Sticks Within 10

Rubric Score: _____ Time Elapsed: _____

Materials: (S) Two linking cube sticks of 5 and one linking cube stick of 7, 9-inch piece of string

1. (Present a 5-stick and the 7-stick.) Compare the length of these two sticks. Use the words *longer than*.
2. Compare the length of one 5-stick to the length of this string. (Show the 9-inch string from Topic A.) Use the words *shorter than*.
3. Break this 5-stick into two parts. Compare the length of this 5-stick (hand student another 5-stick) to the length of the two sticks you are holding now.

What did the student do?	What did the student say?
1.	
2.	
3.	

Module 3: Comparison of Length, Weight, Capacity, and Numbers to 10

A STORY OF UNITS

Mid-Module Assessment Task K•3

Topic C: Comparison of Weight

Rubric Score: _____ Time Elapsed: _____

Materials: (S) Balance scale, pennies, centimeter cubes, 1 light book, 1 heavy book

1. Compare the weight of this book to the weight of this book. Use the words *heavier than*.
2. Put the scissors and the ruler on the balance scale. Use the words *lighter than* to compare their weights.
3. Use the scale to show how many cubes are the same weight as the marker. How many cubes are the same weight as the marker?
4. Use the scale to show how many pennies are the same weight as the marker. How many pennies are the same weight as the marker? Tell me anything else you notice.
5. When we use the words *lighter than* or *heavier than*, what are we comparing?

What did the student do?	What did the student say?
1.	
2.	
3.	
4.	
5.	

Module 3: Comparison of Length, Weight, Capacity, and Numbers to 10

Topic D: Comparison of Volume

Rubric Score: _____ Time Elapsed: _____

Materials: (S) 1 small container (⅛ cup), 1 plastic cup with ½ cup of rice in it, 1 small bowl filled with rice, tub for pouring rice from bowl into cup

1. Compare the capacity of this bowl and this cup. Use the words *more than*. (The student may want to pour to assess or will simply observe to make the comparison.)
2. How many small containers of rice hold the same amount of rice as this large container? (Watch to see what the student does. Ask the student to use the small container to prove his or her answer if the container is not used without prompting.)
3. When we just used the words *more than* or *less than*, what were we comparing?

What did the student do?	What did the student say?
1.	
2.	
3.	

Module 3: Comparison of Length, Weight, Capacity, and Numbers to 10

153

Mid-Module Assessment Task

Standards Addressed — Topics A–D

Describe and compare measurable attributes.

K.MD.1 Describe measurable attributes of objects, such as length or weight. Describe several measurable attributes of a single object.

K.MD.2 Directly compare two objects with a measurable attribute in common, to see which object has "more of"/"less of" the attribute, and describe the difference. *For example, directly compare the heights of two children and describe one child as taller/shorter.*

Evaluating Student Learning Outcomes

A *Progression Toward Mastery* is provided to describe and quantify steps that illuminate the gradually increasing understandings that students develop *on their way to proficiency*. In this chart, this progress is presented from left (Step 1) to right (Step 4). The learning goal for students is to achieve Step 4 mastery. These steps are meant to help teachers and students identify and celebrate what the students CAN do now and what they need to work on next.

Mid-Module Assessment Task K•3

A Progression Toward Mastery

Assessment Task Item and Standards Assessed	STEP 1 Little evidence of reasoning without a correct answer. (1 Point)	STEP 2 Evidence of some reasoning without a correct answer. (2 Points)	STEP 3 Evidence of some reasoning with a correct answer or evidence of solid reasoning with an incorrect answer. (3 Points)	STEP 4 Evidence of solid reasoning with a correct answer. (4 Points)
Topic A K.MD.1 K.MD.2	Student shows little evidence of understanding length comparison.	Student struggles to use the words *longer than* or *shorter than*.	Student may compare the strings well but is unable to perform a small part of the task, for example: ▪ Uses the word *longer than* or *shorter than* incorrectly. ▪ States that the string is being measured rather than the length of the string.	Student: ▪ Says in his or her words that we cannot know because a part is hidden and may mention that the length showing is equal. ▪ Uses the words *longer than* correctly to compare. ▪ Arranges the strings to share an endpoint. ▪ Uses the words *shorter than* correctly to compare. ▪ States that length is being compared or how long the strings are.
Topic B K.MD.2	Student shows little evidence of understanding length comparison.	Student struggles to use the words *longer than* or *shorter than*.	Student demonstrates good understanding of length but may make one small mistake, for example: ▪ Omits or misuses the word *than*.	Student correctly: ▪ Says the 7-stick is longer than the 5-stick. ▪ Says the 5-stick is shorter than the 9-inch string. ▪ Says the two smaller sticks (3-stick and 2-stick or 4-stick and 1-stick) are the same length as the 5-stick.

A Progression Toward Mastery				
Topic C **K.MD.1** **K.MD.2**	Student shows little evidence of understanding of weight.	Student struggles to use the words *heavier than* or *lighter than*. Student may not be sure of how to use the balance.	Student demonstrates good understanding of weight but may make one small mistake, for example: • Omits or misuses the word *than*. • Does not know how to express what is being measured. (This is a challenging generalization and may not come right away.)	Student: • Uses the words *heavier than* correctly to compare. • Uses the words *lighter than* correctly to compare. • Balances the scale with the cubes and says how many cubes are the same as the weight of the marker. • Balances the scale with the pennies and states how many pennies are the same weight as the marker. • States that weight is being compared or how much the books weigh.
Topic D **K.MD.1** **K.MD.2**	Student shows little evidence of understanding of volume.	Student struggles to use the words *more than* or *less than*. Student may not be sure of how to use the containers.	Student demonstrates good understanding but may make one small mistake, for example: • Omits or misuses the word *than*. • Does not know how to express what is being measured. (This is a challenging generalization and may not come right away.)	Student: • Uses the words *more than* correctly to compare. • Measures the rice using the small container and identifies that there are four containers. • States that capacity is being compared or how much the cup holds.

Class Record Sheet of Rubric Scores: Module 3

Name:	Topic A: Comparison of Length and Height	Topic B: Comparison of Length and Height of Linking Cube Sticks Within 10	Topic C: Comparison of Weight	Topic D: Comparison of Volume	Next Steps:

A STORY OF UNITS

Mathematics Curriculum

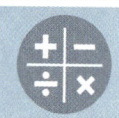

GRADE K • MODULE 3

Topic E
Are There Enough?

K.CC.6

Focus Standard:	K.CC.6	Identify whether the number of objects in one group is greater than, less than, or equal to the number of objects in another group, e.g., by using matching and counting strategies. (Include groups with up to ten objects.)
Instructional Days:	4	
Coherence -Links from:	GPK–M4	Comparison of Length, Weight, Capacity, and Numbers to 5
-Links to:	G1–M3	Ordering and Comparing Length Measurements as Numbers

After the Mid-Module Assessment, the module shifts toward comparison of numbers, opening in Topic E with four lessons in which students consider, "Are there enough?" in a variety of contexts. Students explore and compare area by participating in everyday activities, such as comparing two pieces of paper to determine which one will allow them to create a larger drawing.

In Lesson 16, students consider and establish that a square has enough space to fit a circle inside it and then discover that the same square fits many small squares perfectly if they are arranged in rows.

In Lesson 17, students work to determine whether there are enough forks for every plate, chairs for every child, and pails for every shovel.

In Lessons 18 and 19, the language of *enough* shifts to the language of *more than* and *fewer than*. There are more forks than plates. There are fewer chairs than children. There are the same number of pails and shovels.

Topic E

A Teaching Sequence Toward Mastery of Are There Enough?

Objective 1: Make informal comparison of area.
(Lesson 16)

Objective 2: Compare to find if there are enough.
(Lesson 17)

Objective 3: Compare using *more than* and *the same as*.
(Lesson 18)

Objective 4: Compare using *fewer than* and *the same as*.
(Lesson 19)

Lesson 16

Objective: Make informal comparison of area.

Suggested Lesson Structure

- ■ Fluency Practice   (13 minutes)
- ■ Application Problem   (5 minutes)
- ■ Concept Development   (25 minutes)
- ■ Student Debrief   (7 minutes)
- **Total Time**   **(50 minutes)**

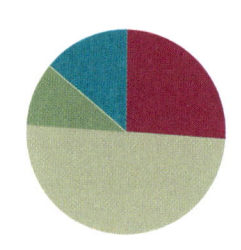

Fluency Practice (13 minutes)

- Groups of Shapes **K.G.2**   (5 minutes)
- Show Me Bigger and Smaller **K.MD.1**   (3 minutes)
- Building Up to the Sprint Routine: Starting and Stopping at the Signal **K.CC.3**   (5 minutes)

Groups of Shapes (5 minutes)

Materials: (T) Shape signs (Fluency Template 1), music (S) Shape cutouts (include exemplars and non-examples) (Fluency Template 2)

Note: This activity prepares students for the current lesson by providing a quick review of shapes.

- T: Choose a shape, and then meet me at the rug.
- T: Look at your shape. Raise your hand if you know the name of your shape. When I give the signal, whisper the name of your shape to yourself. Ready? (Signal.)
- S: (Whisper the shape name.)
- T: Look around the room. Do you see signs with pictures of shapes?
- S: Yes.
- T: Do you see your shape?
- S: Yes.
- T: When I start the music, I want you to calmly walk to the sign that has the same shape as yours.
- T: When I point to your group, say the name of your shape. (Point to the group of triangles.)
- S: Triangles!

Continue identifying the remaining groups, and then call students back to the rug to trade for a new shape. Circulate to observe which students struggle with this task, and provide support by having them identify the attributes of their shape as compared to the shapes pictured on the signs.

Show Me Bigger and Smaller (3 minutes)

Note: This activity prepares students for the current lesson by making visual and kinesthetic connections to size comparison.

Conduct similarly to the Show Me Taller and Shorter activity in Lesson 2, but have students position their hands close together as if holding a tennis ball to indicate *smaller* and hands farther apart as if holding a basketball to indicate *bigger*.

Building Up to the Sprint Routine: Starting and Stopping at the Signal (5 minutes)

Materials: (S) Lined writing paper

Note: Although the task is simple, this activity conditions students to stop working, even when they have not finished; additionally, it develops the self-regulation necessary for participating in Sprints. Teaching the Sprint routine in stages may be time-consuming, but the investment is worthwhile. Students begin their first Sprint in Lesson 21.

- T: When I say "go," we are going to practice writing numbers 1–10 quickly but carefully, like this. (Demonstrate.) When you hear the bell ring, you must stop and hold up your pencil, even if you are not finished. What do you do when you hear the bell?
- S: Stop and hold up my pencil.
- T: Good. Remember, it's okay if you don't finish. Ready? Go!
- S: (Write numbers 1–10.)
- T: (Before students reach 10, ring the bell.) Pencils up, up, up!
- S: (Hold pencils up.)
- T: Wow! You really followed the directions! Let's practice again. Ready? Go!

Continue several more times, praising students for following directions rather than completing the task.

Application Problem (5 minutes)

Materials: (S) Playing card, bag of linking cubes

How many linking cubes would you need to cover up your card? Make a guess! Now, work with your partner to test your guess. What did you discover? How many cubes did you need? Did your friends use the same number of cubes?

Note: This problem requires students to compare the area of the face of the linking cube to the area of the rectangular card. This sets the stage for today's lesson. Circulate during the exercise to determine which face of the cube the students choose to use. Observe whether they choose to stack the cubes, as well as whether they arrange them without gaps.

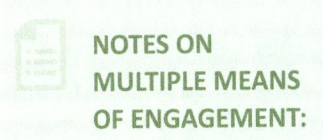

NOTES ON MULTIPLE MEANS OF ENGAGEMENT:

Support English language learners by providing them with sentence frames such as, "To cover my card, I think I will need ___ linking cubes." and "To cover my card, I needed ___ linking cubes." This makes it possible for English language learners to benefit from the Application Problem and discussion with a partner. For non-readers, these frames can be communicated orally.

A STORY OF UNITS   Lesson 16   K•3

Concept Development (25 minutes)

Materials: (T) 1 set of student materials (S) My square recording sheet (Template), 1 four-inch square of construction paper, 1 four-inch diameter paper circle, 20 one-inch paper or plastic square tiles, 1 small bag of large flat beans, my square recording sheet (Template)

MP.7

T: Place your square of paper on your desk. What are some things that would fit onto your square?
S: My hand! → An apple. → A block. → Some crayons.
T: Will there be enough room for a circle like this? (Hold up a circle.) (Various responses.) Let me give each of you a circle to test your guess.
S: Yes! It fits.
T: Will you have enough room for another circle?
S: No.
T: On your recording sheet, let's draw what your square looks like now. (Demonstrate.)
T: (Hold up a 1-inch paper square.) Will this fit on your paper? Is there enough space?
S: Yes.
T: Do you think 5 of them will fit? (Various responses.) Take out your paper squares, and put 5 of them on the paper. Is there enough space for 1 more?
S: Yes!
T: Put another square on the paper. (Repeat until the square is filled with smaller squares. Notice student strategies as they try to fit more onto their paper.)
T: So, your square held 1 circle. How many small squares did you need to fill your big square?
S: 16.
T: Let's draw what we did on the recording sheet. (Demonstrate.)
T: I wonder how many beans you would need to cover your square? (Various responses.) Work with your partner to put as many beans as you can on your square without piling them. (Allow time for experimentation and discussion.)
T: What did you notice about using the beans?
S: It was harder! → They didn't fit together like the squares did. → I can still see some of the square. → It took a lot of them!
T: Did you use more beans or squares in this activity?
S: More beans! → We had to use more because they were smaller.

> **NOTES ON MULTIPLE MEANS OF ACTION AND EXPRESSION:**
>
> Students may need to use more beans than previously counted in class. Some students might use this opportunity to showcase their knowledge of counting beyond 20. Other students might benefit from counting beyond 20 with support. Knowing how many more is not the objective in this lesson. Simply knowing that there are more beans than squares is sufficient for an informal comparison.

> **NOTES ON MULTIPLE MEANS OF ACTION AND EXPRESSION:**
>
> Challenge students working above grade level to explain how they were able to cover one square with one circle, 16 small squares, or many beans. Encourage them to use the math words that they know.

Lesson 16: Make informal comparison of area.

Problem Set (10 minutes)

Students should do their personal best to complete the Problem Set within the allotted time.

Student Debrief (7 minutes)

Lesson Objective: Make informal comparison of area.

The Student Debrief is intended to invite reflection and active processing of the total lesson experience.

Invite students to review their solutions for the Problem Set. They should check work by comparing answers with a partner before going over answers as a class. Look for misconceptions or misunderstandings that can be addressed in the Debrief. Guide students in a conversation to debrief the Problem Set and process the lesson.

Any combination of the questions below may be used to lead the discussion.

- Were you able to cover the square entirely with your little squares or the beans? Why?
- Was that also true on the recording sheet? How were the two different? (Listen for discussion about how differences in sides and angles affected their work. Some students may notice the space between smaller units. Do not worry if they do not notice this because it is a concept they will encounter in Grade 3.)
- What strategies did you use to fit more things onto your paper?
- When you were covering the square, how did you decide that you were done? When you were covering the shapes on the Problem Set, how did you decide that you were done?
- Were you surprised by the number of squares or beans needed to cover some of the shapes?
- What math vocabulary did we use today to communicate precisely?

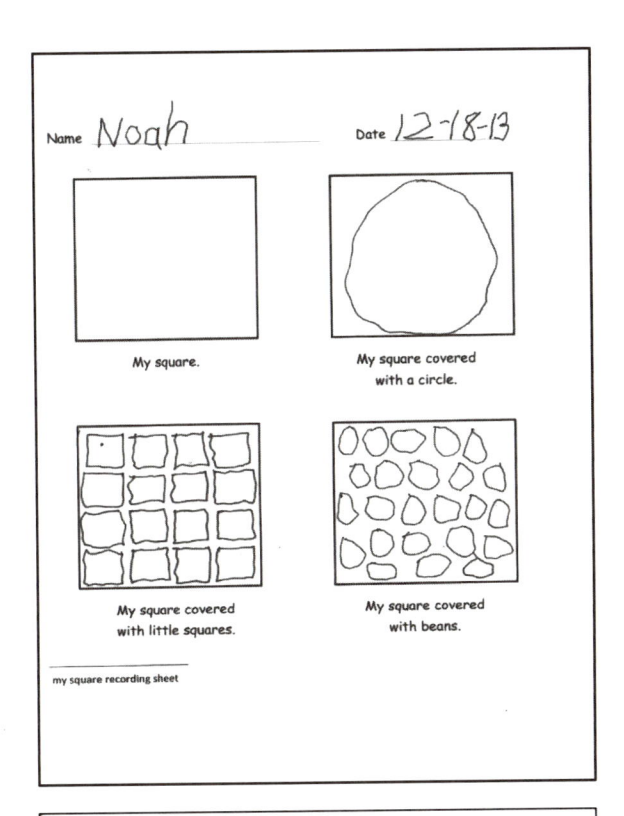

A STORY OF UNITS

Lesson 16 Problem Set K•3

Name _____ Date _____

Cover the shape with squares. Count how many, and write the number in the box.

☐ Squares

Cover the shape with beans. Count how many, and write the number in the box.

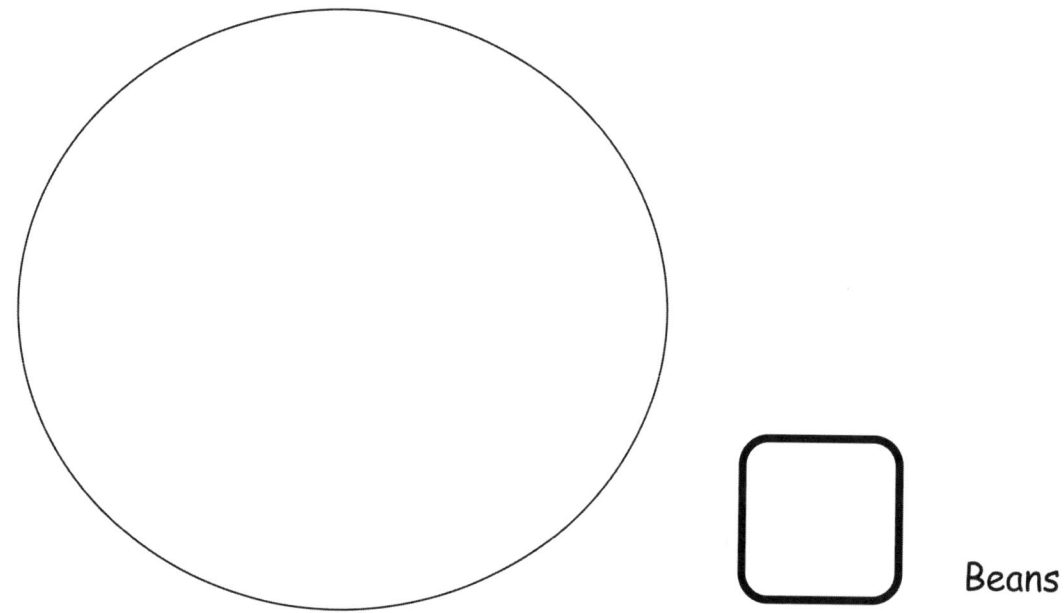

☐ Beans

Lesson 16: Make informal comparison of area.

Lesson 16 Homework K•3

Name _____ Date _____

Trace your hand. Cover the tracing with pennies. Have an adult trace his or her hand. Cover the tracing with pennies.* Whose hand is bigger? How do you know that?

*Note: Instead of pennies, you can use pasta, beans, buttons, or another coin. You may want to do this activity twice using different materials to cover the hands. Talk about which materials took more or less to cover and why.

Triangle

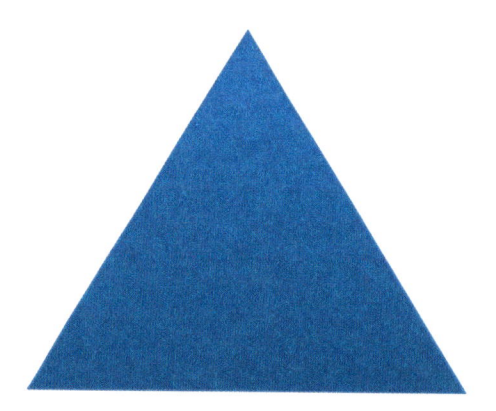

Rectangle

shape signs

Square

Hexagon

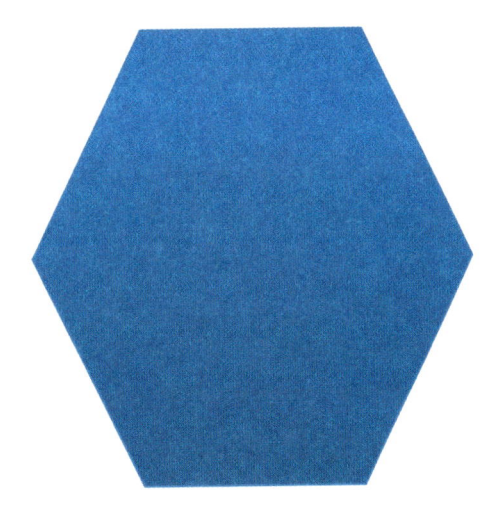

shape signs

Circle

shape signs

A STORY OF UNITS Lesson 16 Fluency Template 2 K•3

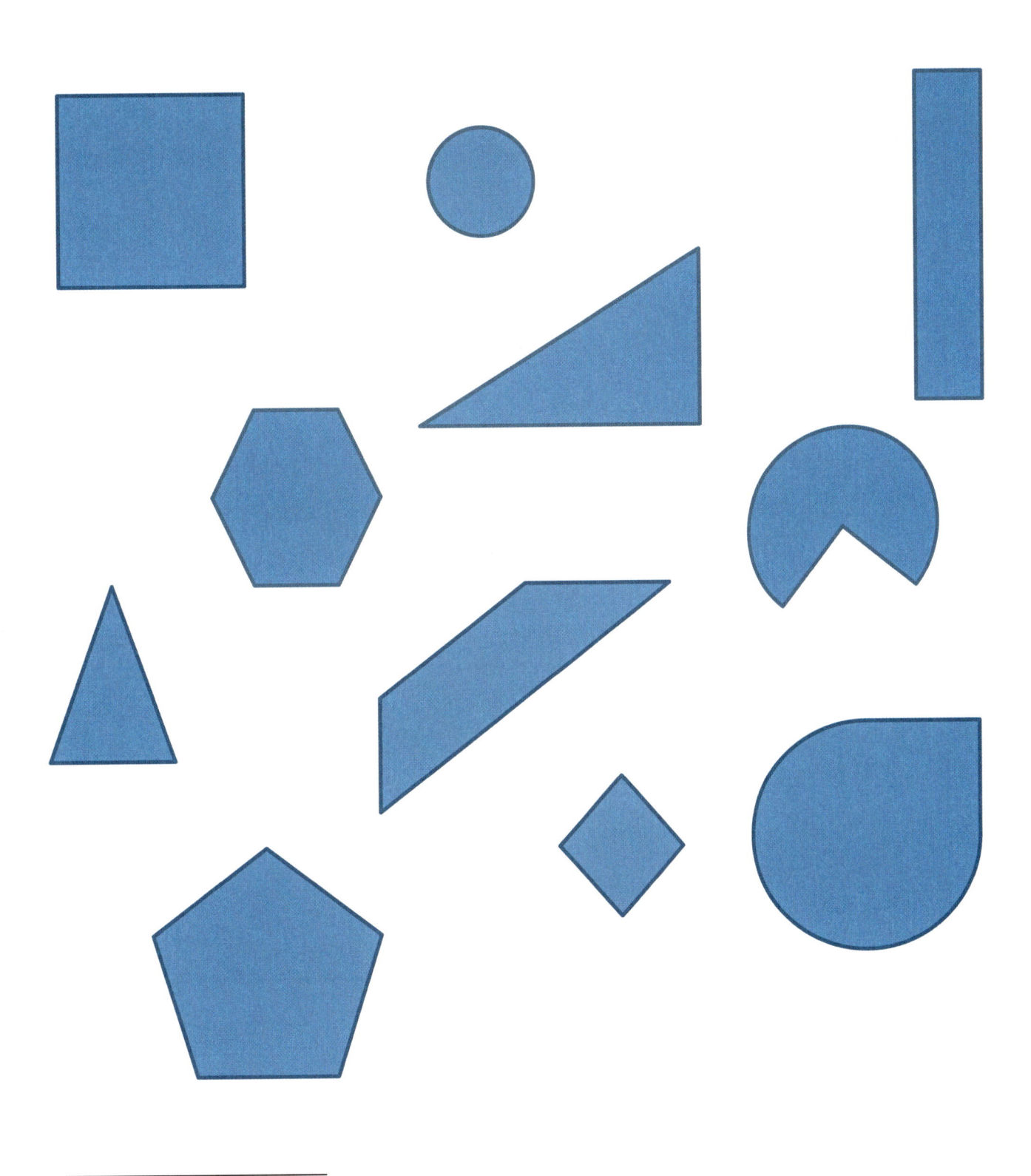

shape cutouts

Lesson 16: Make informal comparison of area.

Name _____ Date _____

My square.

My square covered with a circle.

My square covered with little squares.

My square covered with beans.

my square recording sheet

Lesson 16: Make informal comparison of area.

A STORY OF UNITS   Lesson 17   K•3

Lesson 17

Objective: Compare to find if there are enough.

Suggested Lesson Structure

- ■ Fluency Practice   (11 minutes)
- ■ Application Problem   (8 minutes)
- ■ Concept Development   (25 minutes)
- ■ Student Debrief   (6 minutes)
- **Total Time**   **(50 minutes)**

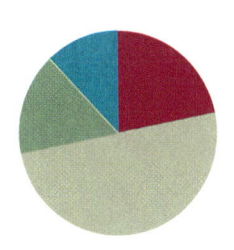

Fluency Practice (11 minutes)

- Dot Cards of 8 **K.CC.5, K.CC.2**   (4 minutes)
- Show Me Bigger and Smaller **K.MD.1**   (3 minutes)
- Matching Fingertips One-to-One **K.CC.6** (4 minutes)

Dot Cards of 8 (4 minutes)

Materials: (T/S) Dot cards of 8 (Fluency Template)

Note: This activity deepens students' knowledge of embedded numbers and develops part–whole thinking, which is foundational to the work of upcoming modules.

　　T: (Show a card with 8 dots.) How many dots do you count? Wait for the signal to tell me.
　　S: 8.
　　T: How can you see them in two parts?
　　S: (Point to the card.) I saw 4 here and 4 here. → I saw 5 here and 3 here. → I saw 6 here and 2 here.

Repeat with other cards. Pass out the cards for students to work independently.

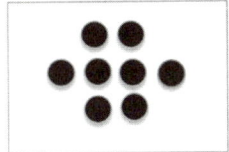

Show Me Bigger and Smaller (3 minutes)

Note: This activity prepares students for the current lesson by making visual and kinesthetic connections to size comparison. Conduct the activity similarly to the Show Me Taller and Shorter activity in Lesson 2, but have students position their hands close together as if holding a tennis ball to indicate *smaller* and hands farther apart as if holding a basketball to indicate *bigger*.

Lesson 17: Compare to find if there are enough.   171

Matching Fingertips One-to-One (4 minutes)

Materials: (S) Dice

Note: This exercise relates to the concept of *enough* and anticipates drawing lines to match one-to-one pictorially in upcoming lessons.

1. Partner A rolls a die and shows as many fingers as dots on the rolled die.
2. Partner B shows the same number of fingers.
3. Both partners touch fingertips, carefully matching one-to-one.

Application Problem (8 minutes)

Materials: (T) Music player; chairs, carpet squares, or pieces of construction paper per student; plus several more chairs than students

It's time to have a math celebration and play a game of musical chairs (or carpet squares or papers)! During the first round, make sure that there are several more chairs than students. When the students sit and notice the extra chairs, tell them, "There are not enough children to fill the chairs." Continue playing and remove a chair each round until there are just as many chairs as students. When they sit down, tell them, "There are just enough chairs!" Repeat as time permits.

Note: The physical matching of chairs to students and the introduction of *enough* serves as the anticipatory set for the lesson.

Concept Development (25 minutes)

Materials: (S) Paper plate, cup, spoon, and napkin; popcorn (or some other snack); bottle of water

Setup: Arrange students into groups of four around tables or on the floor.

MP.6

T: We are going to have a math popcorn party today! Let's set our tables. You will each need a plate, a spoon, a cup, and a napkin. Here is a plate for each of you. (Hand each group of students a stack of four plates.) Please pass out the plates. Are there **enough**?

S: Yes.

T: Good! There is one plate for each of you.

T: Here are spoons for you. (Hand each group of 4 students three spoons.) Are there enough spoons?

S: There are **not enough**!

T: How many more do you need in your group?

S: We need one more.

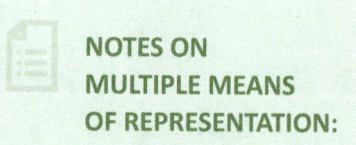

NOTES ON MULTIPLE MEANS OF REPRESENTATION:

Scaffold the lesson for English language learners by holding up a plate, spoon, cup, and napkin while explaining what each person in the group needs before handing out the materials to the groups to discover whether they have enough.

A STORY OF UNITS — Lesson 17   K•3

T: (Give each group one more spoon.)
T: Now, we have one spoon for each child. We have enough spoons.
T: Here are your cups. (Give each group five cups.)
S: There are too many!
T: How many extra cups do you have in your group?
S: One.
T: (Take back the extra cups.)
T: Good! Each of you has a cup. We have **just enough**.
T: Here are some napkins for you. Make sure that you each have one. (Hand each group two napkins.)
S: There are not enough! We need more!
T: Here are extra napkins. Please make sure that each of you has one napkin. When you have enough, please give the rest back to me. Also, here is some popcorn for you to munch on while you do your Problem Set! (Serve the popcorn and water.) I wonder how many cups I can fill with my water bottle? I hope that I have enough!

MP.6

> **NOTES ON MULTIPLE MEANS OF REPRESENTATION:**
>
> For students who struggle, clarify what the terms *just enough* and *not enough* mean after they are introduced. Provide examples of *just enough* while asking if there are any left over, and provide examples of *not enough* while asking if everyone received one.

Have the students count the cups the Say Ten way as you pour. (Support the students as they count above 20.) Use a scoop to distribute popcorn, and count the number of scoops. Discuss with the class what you should do with the leftovers (the remainder).

Problem Set (10 minutes)

Students should do their personal best to complete the Problem Set within the allotted time.

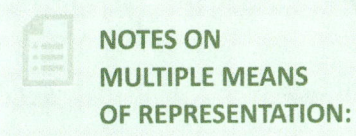

Lesson 17: Compare to find if there are enough.

EUREKA MATH

Student Debrief (6 minutes)

Lesson Objective: Compare to find if there are enough.

The Student Debrief is intended to invite reflection and active processing of the total lesson experience.

Invite students to review their solutions for the Problem Set. They should check work by comparing answers with a partner before going over answers as a class. Look for misconceptions or misunderstandings that can be addressed in the Debrief. Guide students in a conversation to debrief the Problem Set and process the lesson.

Any combination of the questions below may be used to lead the discussion.

- When we were playing musical chairs, did we know before we started if there would be enough chairs?
- How could we have found out if there were enough chairs before we started playing?
- When there were **not enough** spoons, how did you know how many more your group needed?
- When there were too many cups, how did you know how many extra cups your group had?
- In the Problem Set, were there **just enough** flowers for the butterflies? How did you know?
- How many plates did you draw on the back of your paper? How many apples did you draw? Did you draw enough apples and plates?
- What new (or significant) math vocabulary did we use today to communicate precisely?

A STORY OF UNITS

Lesson 17 Problem Set   K•3

Name _____   Date _____

Draw straight lines with your ruler to see if there are enough flowers for the butterflies.

On the back, draw some plates. Draw enough apples so each plate has one.

Lesson 17:   Compare to find if there are enough.

175

A STORY OF UNITS

Lesson 17 Homework K•3

Name _____ Date _____

Draw straight lines with your ruler to see if there are enough shovels for the pails.

Lesson 17: Compare to find if there are enough.

A STORY OF UNITS

Lesson 17 Homework K•3

Make sure there is a fork for every plate. Draw straight lines with a ruler from each plate to a fork. If there are not enough forks, draw one.

You have 4 fishes. Draw enough fish bowls so you can put 1 fish in each fish bowl.

Lesson 17: Compare to find if there are enough.

177

dot cards of 8

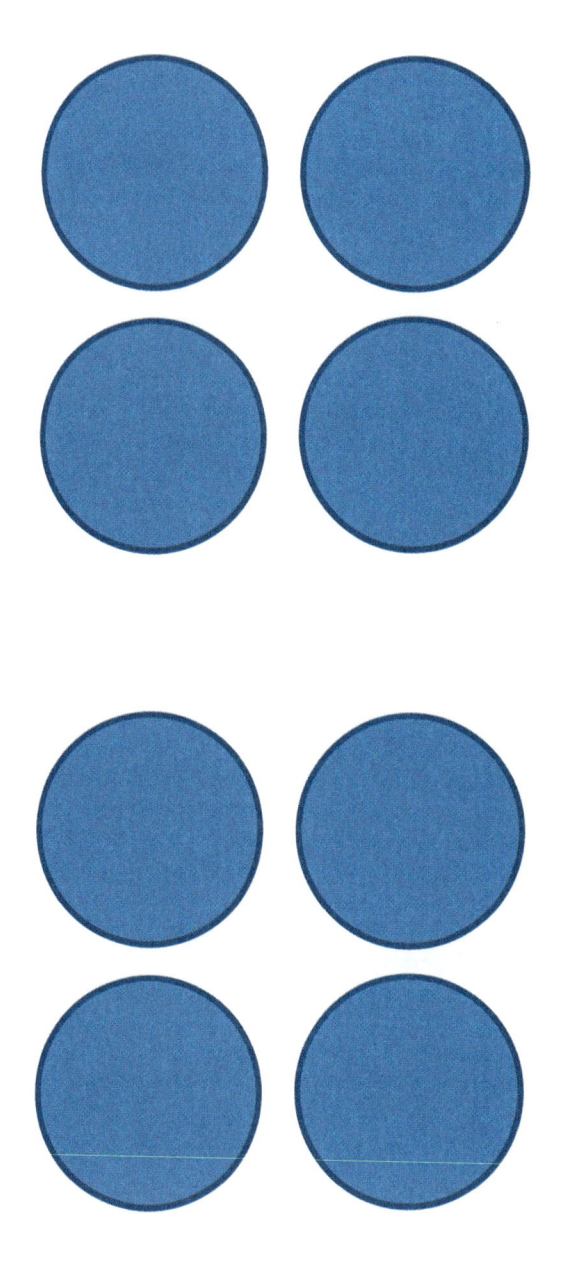

dot cards of 8

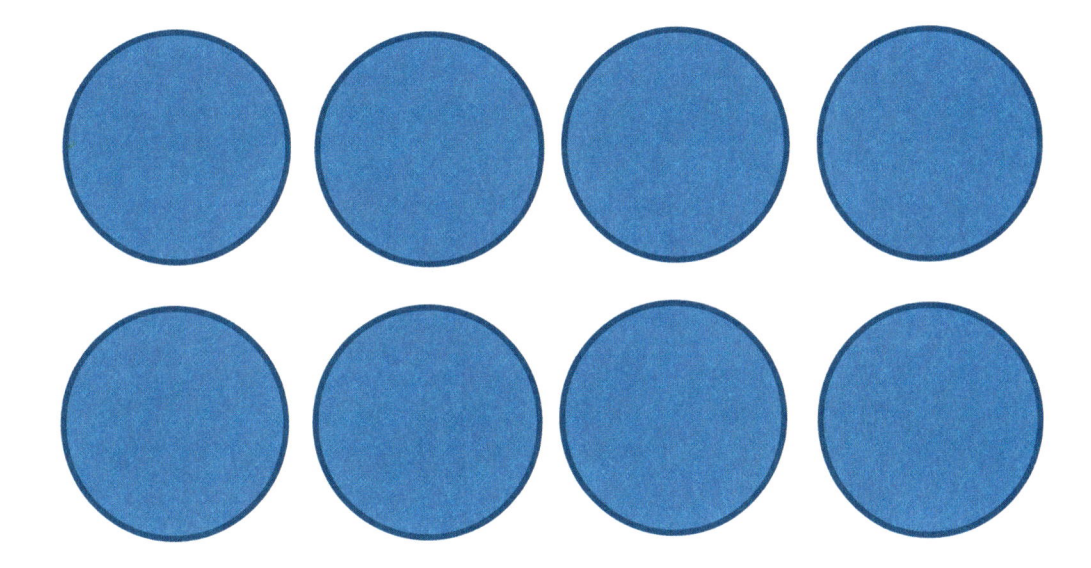

dot cards of 8

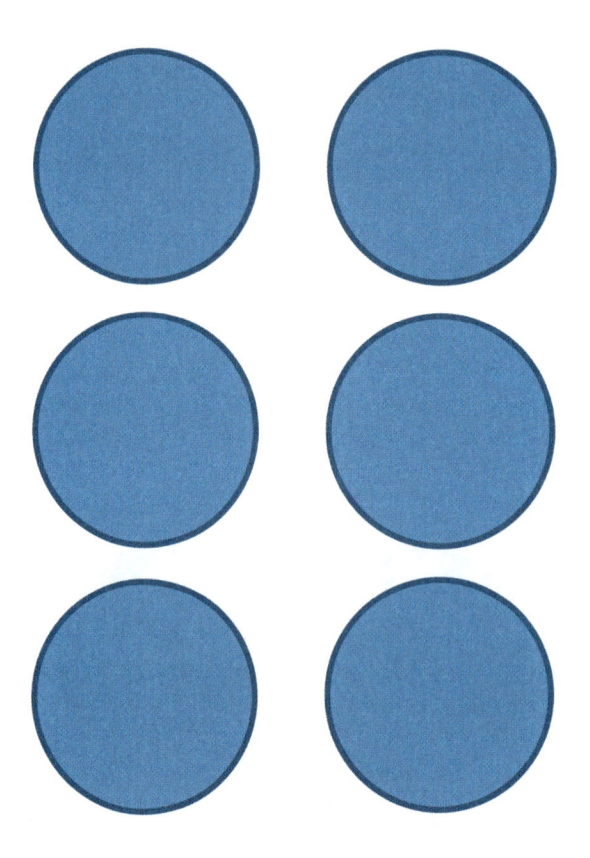

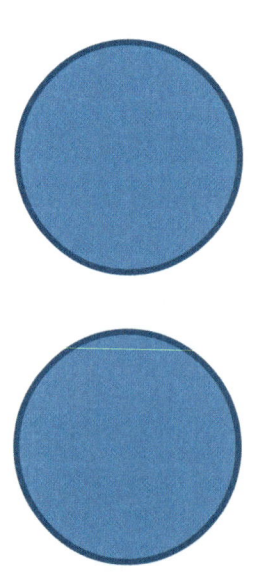

dot cards of 8

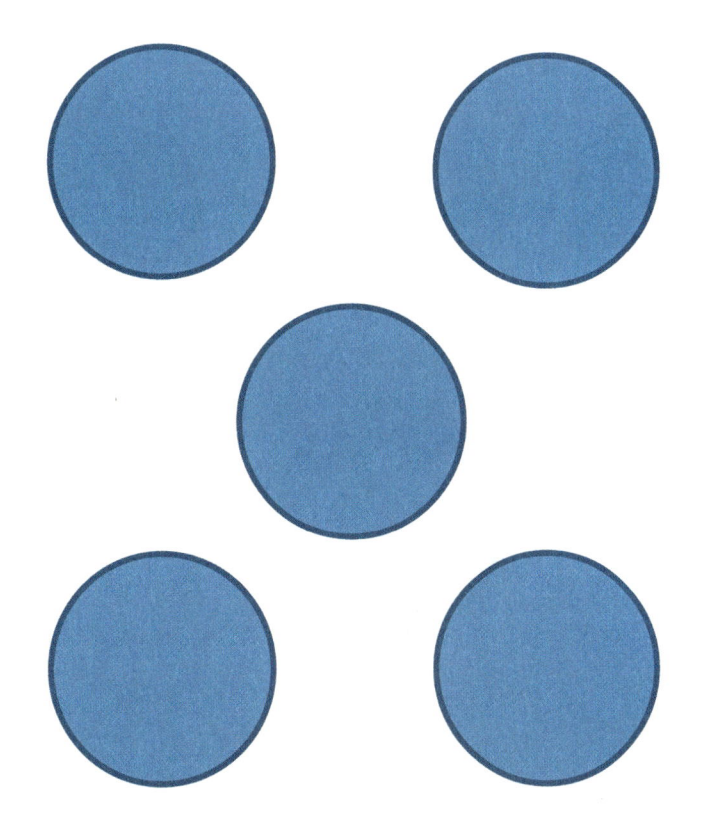

dot cards of 8

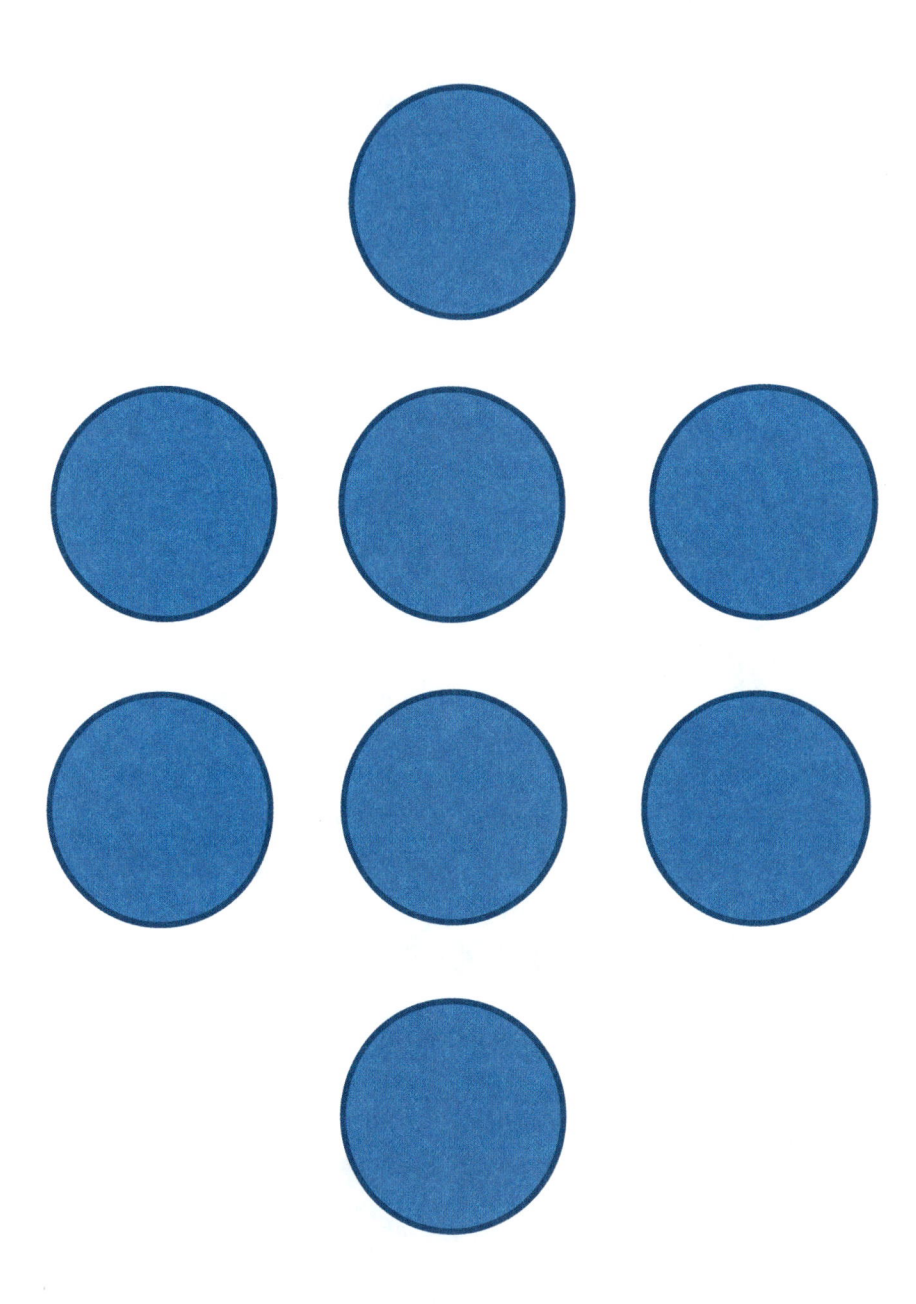

dot cards of 8

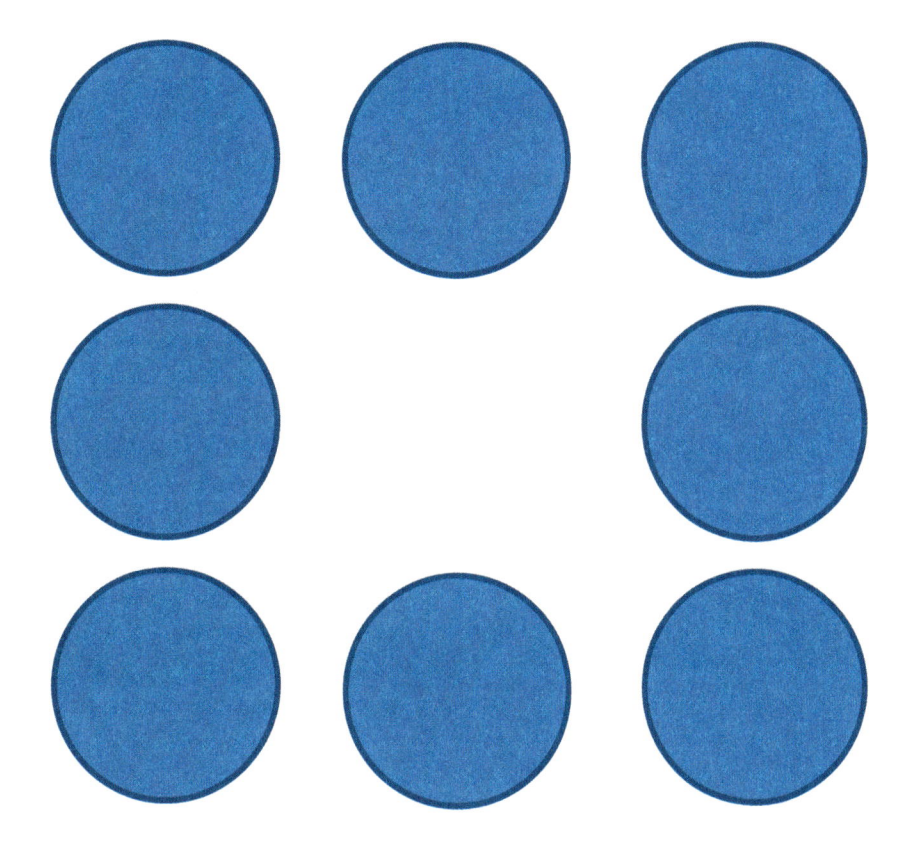

A STORY OF UNITS — Lesson 17 Fluency Template — K•3

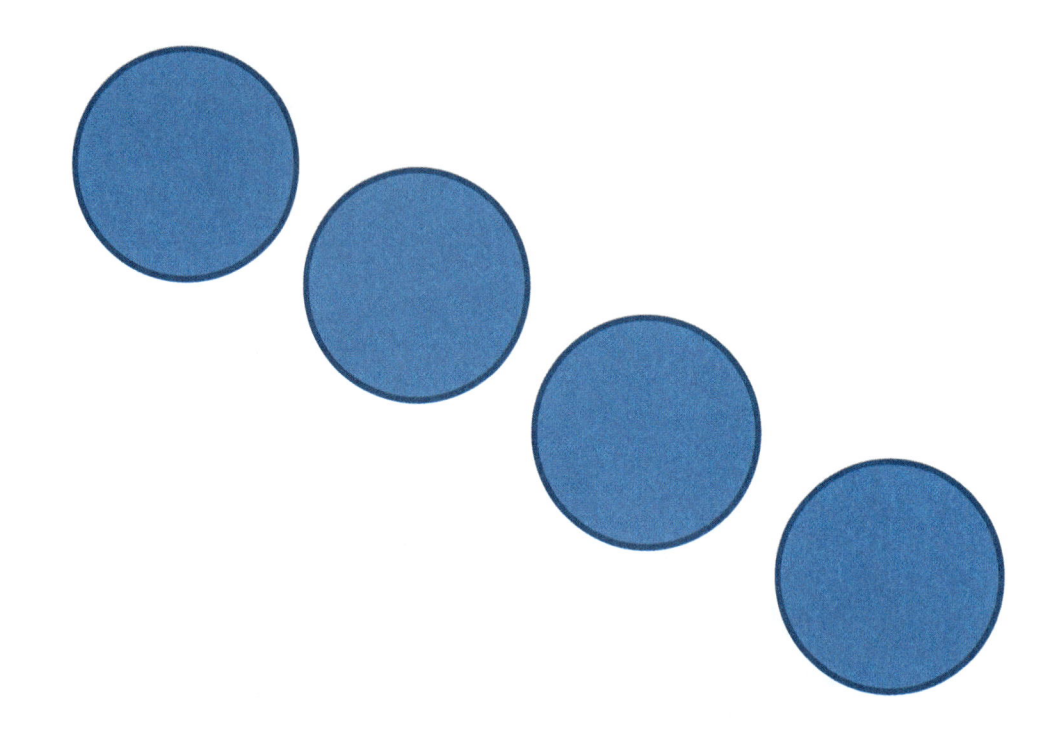

dot cards of 8

184 Lesson 17: Compare to find if there are enough.

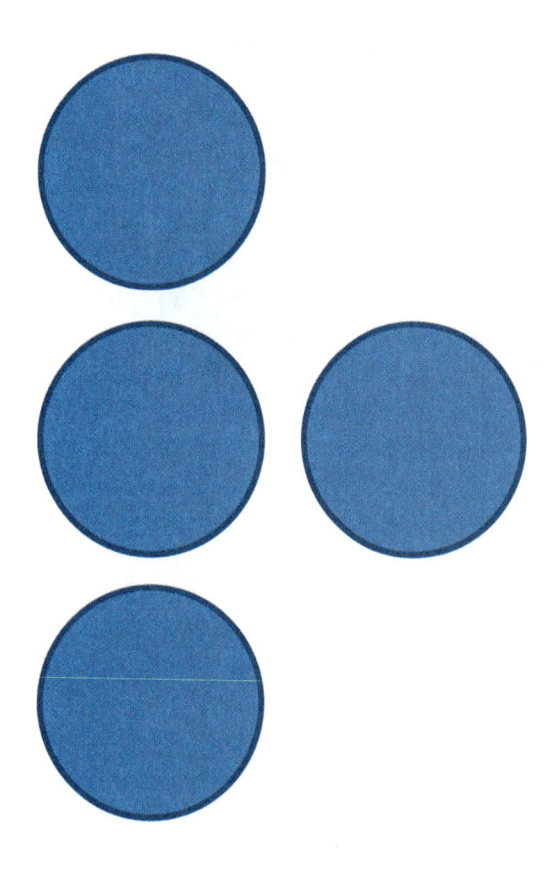

© 2015 Great Minds. eureka-math.org
GK-M3-TE-B3-1.3.1-01.2016

EUREKA MATH

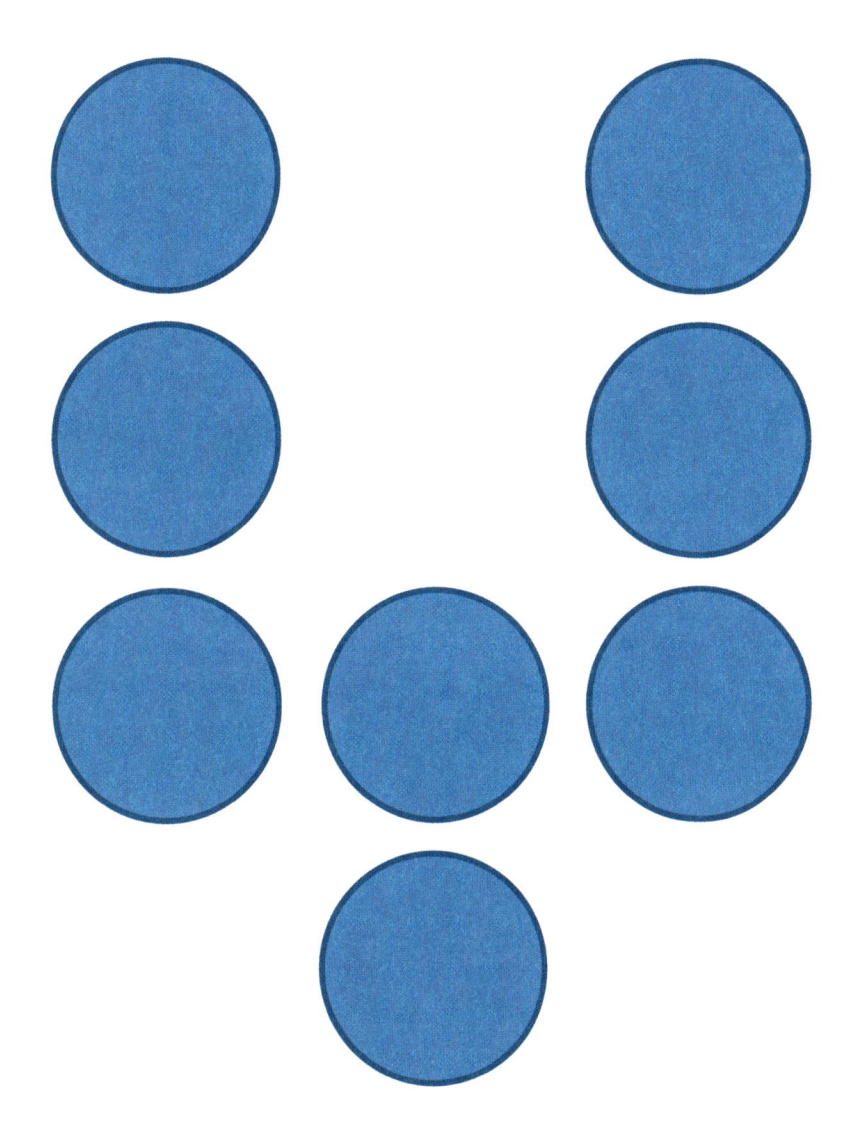

dot cards of 8

Lesson 17: Compare to find if there are enough.

dot cards of 8

Lesson 18

Objective: Compare using *more than* and *the same as*.

Suggested Lesson Structure

- **Fluency Practice** (12 minutes)
- **Application Problem** (5 minutes)
- **Concept Development** (25 minutes)
- **Student Debrief** (8 minutes)
- **Total Time** **(50 minutes)**

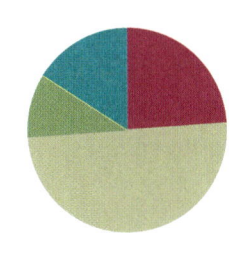

Fluency Practice (12 minutes)

- Finger Number Pairs **K.CC.4a** (4 minutes)
- Matching Fingertips One-to-One **K.CC.6** (4 minutes)
- Matching Circles and Squares **K.CC.6** (4 minutes)

Finger Number Pairs (4 minutes)

Note: This activity ensures that students do not become overly reliant on counting the Math Way and gives them yet another method of breaking apart numbers, which is essential to the next module's work.

- T: You've gotten very good at showing fingers the Math Way. I want to challenge you to think of other ways to show numbers on your fingers. Here's a hint: you can use two hands! First, I'll ask you to show me fingers the Math Way. Then, I'll ask you to show me the number another way. Ready? Show me 2.
- S: (Hold up the pinky and ring fingers of the left hand.)
- T: Now, show me another way to make 2 using two hands.
- S: (Show 1 finger on each hand.)
- T: How can we be sure that we're still showing 2?
- S: Count the fingers on both hands.

Continue the process with other numbers. For numbers where more than one combination is possible, have students try each other's combinations.

Matching Fingertips One-to-One (4 minutes)

Note: This exercise allows students to demonstrate the concepts of *enough* and *same as;* it also anticipates drawing lines to match one-to-one for comparison in upcoming lessons.

Conduct the activity as described in Lesson 17, but now, invite students to show fingers in a variety of ways and verify that it is still the same number of fingers.

A STORY OF UNITS — Lesson 18 K•3

Matching Circles and Squares (4 minutes)

Materials: (S) Dice, personal white board

Note: Students gain experience with equivalency and practice one-to-one matching in anticipation of comparison.

1. Partner A rolls a die and draws the number of circles that corresponds to the number of dots on the rolled die.
2. Partner B draws that same number of squares.
3. Partner A draws lines to match circles to squares, while both partners say, "One circle, one square, one circle, one square…."

Application Problem (5 minutes)

Draw four little mice. Draw some pieces of cheese so that each mouse can have one. Use a ruler to draw a line between each mouse and its cheese. Are there just enough pieces of cheese? Talk to your partner about how you knew how many pieces of cheese to draw.

Note: Circulate during the activity to check understanding of one-to-one correspondence before today's lesson.

Concept Development (25 minutes)

Materials: (T) Basket of 3 blocks or small toys, additional blocks (S) Bag of 5 loose red linking cubes, bag of 10 loose blue linking cubes, pair of dice with the 6-dot side covered, 5 additional red linking cubes

MP.7

T: (Call four students to the front.) Please reach in, one at a time, and take one thing out of my mystery basket.

S: (Last student reaches inside.) I don't have one!

T: There are not enough. There are more students than blocks. Here is another block for you to hold. Now, we have the same number of blocks as students! Please return to your seats.

T: What happened when I asked them each to take a block?

S: There weren't enough.

T: Right. I had more students than blocks! I had to find another block to make them the same number. (Put another three blocks in the basket, and call a pair of students forward.) One at a time, please take one thing out of my basket. (Show students the remaining block in the basket.) What happened this time?

S: There were too many!

T: There were more blocks than students! Student A, would you please come up and take a block out of my basket? (Student takes last block.) Now, we have the same number of blocks as students.

Repeat the activity several times until many or all students have had a chance to participate. Model and encourage use of "more ____ than ____" and "the same number of ____ as ____."

---

188   Lesson 18:   Compare using *more than* and *the same as*.

EUREKA MATH

T: Take out your bags of linking cubes. Put the red cubes on one side of your desk and the blue cubes on the other side of your desk. Take a minute to look at the cubes. Tell your partner what you notice. Are the red and blue cube sets the same? (Allow time for discussion. Circulate to notice how students compare sets. Do they make towers out of the cubes? Do they just line them up and notice what is missing? Do they pair them and see what is left? Do they count them?)

T: What do you notice?

S: There are more blue cubes.

T: Tell me how you knew.

S: I lined them up like this and saw that this line was longer. → I made towers out of my cubes, and this tower was higher. → I counted five here and ten here. → I could just see without counting.

T: Those are interesting strategies! You found ways to know that there were more blue cubes than red cubes. Now, put seven blue cubes back into your bag. What do you notice about the cubes you have left on your desk?

S: Now, there are more red ones!

T: Yes, now there are more red cubes than blue cubes. Can you put enough red cubes away so that there are the same number of red cubes as blue cubes? Show your work to your partner. (Circulate again to ensure understanding of one-to-one correspondence.)

Invite students to play a game. After distributing extra materials (the dice and additional red linking cubes), have one student roll a pair of dice, and then show the same number of red cubes as the number rolled. The partner does the same. Demonstrate how to make a *more than* or *the same as* statement based on what happened. Circulate and support students as they practice making precise math statements.

Problem Set (10 minutes)

Students should do their personal best to complete the Problem Set within the allotted time.

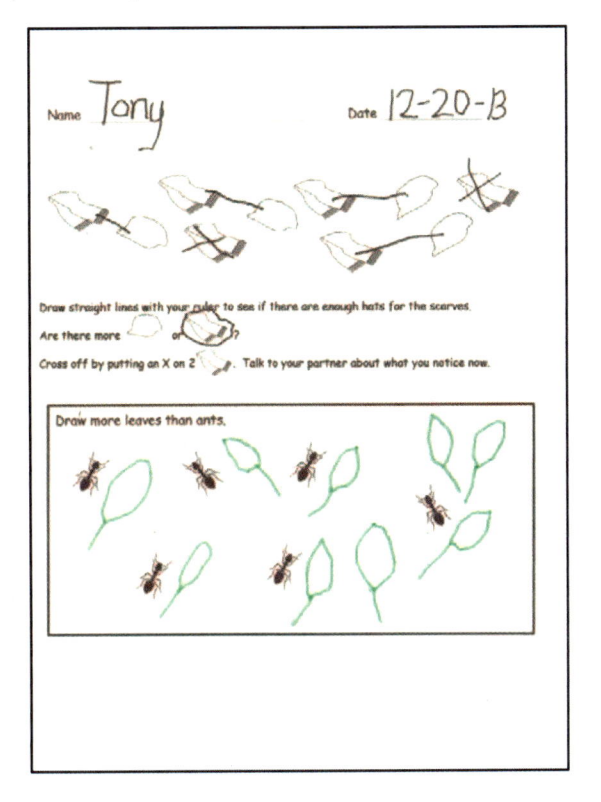

Lesson 18: Compare using *more than* and *the same as*.

Student Debrief (8 minutes)

Lesson Objective: Compare using *more than* and *the same as*.

The Student Debrief is intended to invite reflection and active processing of the total lesson experience.

Invite students to review their solutions for the Problem Set. They should check work by comparing answers with a partner before going over answers as a class. Look for misconceptions or misunderstandings that can be addressed in the Debrief. Guide students in a conversation to debrief the Problem Set and process the lesson.

Any combination of the questions below may be used to lead the discussion.

- What happened when you first took out the red and blue cubes? How did you know which set had more? Did someone else do it differently?
- In the Problem Set, were there more hats or scarves? How did you know? (Help students use **more than** and **the same** number **as** in their answers.)
- How did you use your ruler to help find which had more?
- What happened when you crossed out the two scarves? (Guide students to practice saying *more than* and the *same as*.)
- How many ants were there? You had to draw more leaves than ants. How many leaves did you draw? Check with your partner to see if they drew the same number of leaves. Who had more?
- What new math vocabulary did we use today to communicate precisely?

A STORY OF UNITS　　　　　　　　　　　　　　　　　　　Lesson 18 Problem Set   K•3

Name _____   Date _____

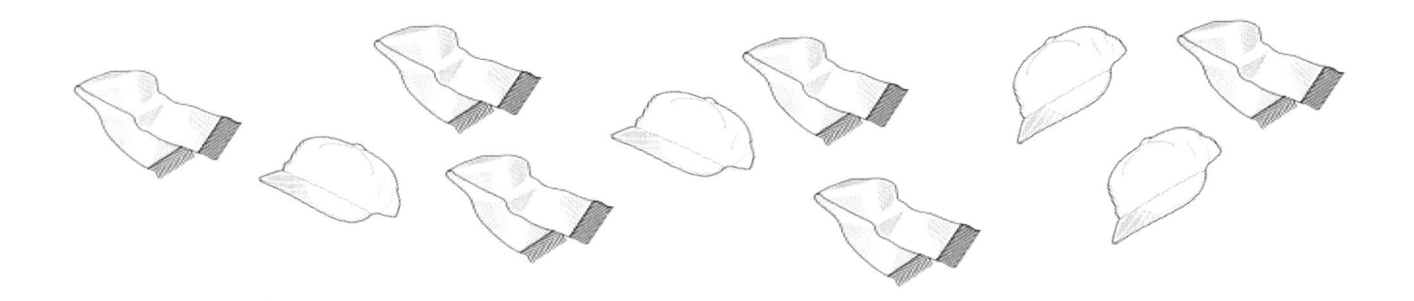

Draw straight lines with your ruler to see if there are enough hats for the scarves.

Are there more ⌢ or ⌁ ?

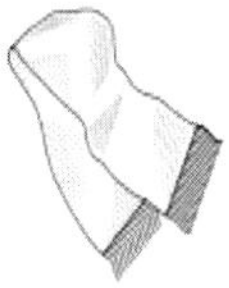

Cross off by putting an X on 2 ⌁. Talk to your partner about what you notice now.

Draw more leaves than ants.

Lesson 18:   Compare using *more than* and *the same as*.

191

Name _____ Date _____

Draw straight lines with your ruler to see if there is one hoop for each ball.

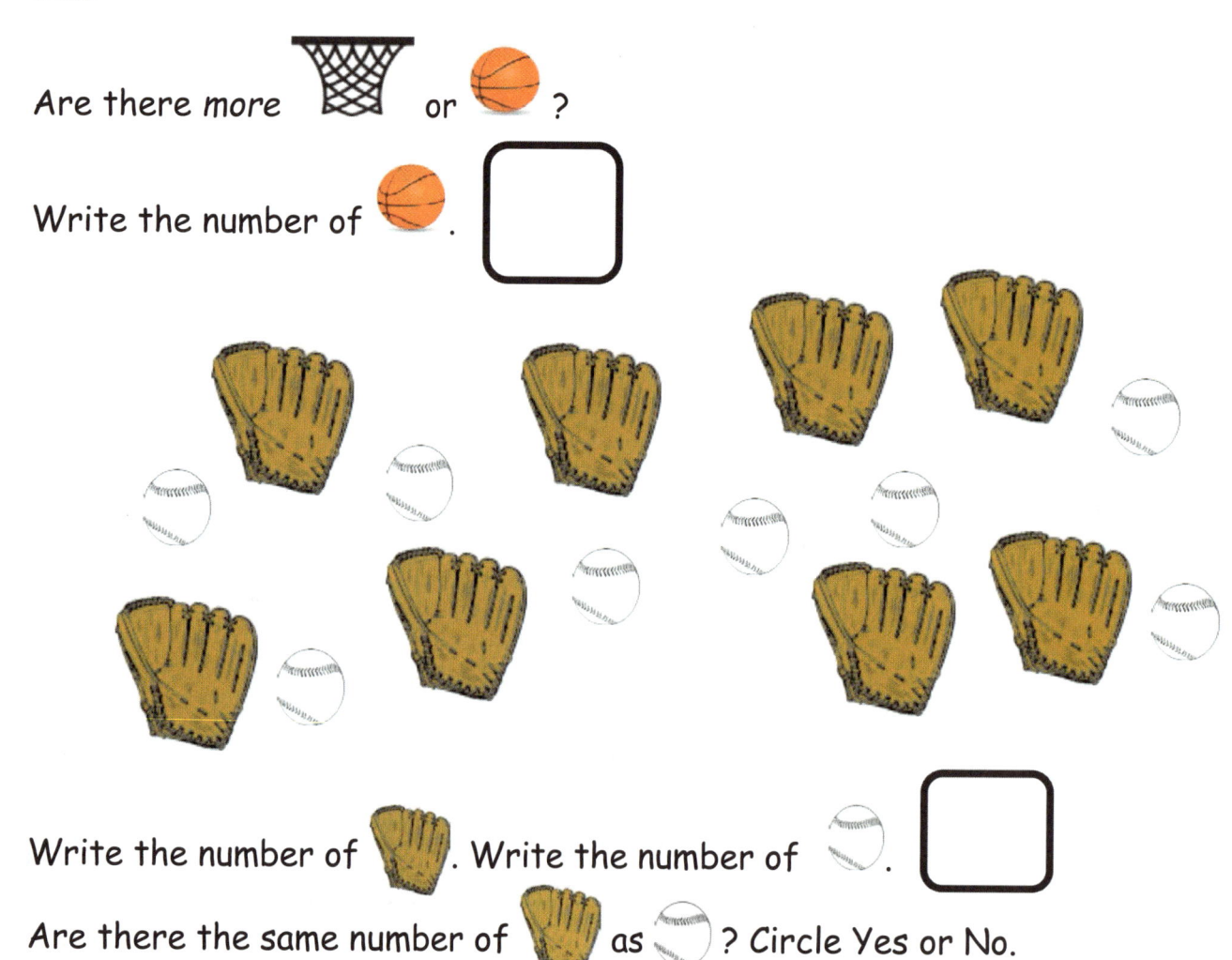

Are there *more* 🏀 or 🏀 ?

Write the number of 🏀. ☐

Write the number of 🥎. Write the number of ⚾. ☐

Are there the same number of 🥎 as ⚾? Circle Yes or No.

Lesson 18: Compare using *more than* and *the same as*.

A STORY OF UNITS — Lesson 19 K•3

Lesson 19

Objective: Compare using *fewer than* and *the same as*.

Suggested Lesson Structure

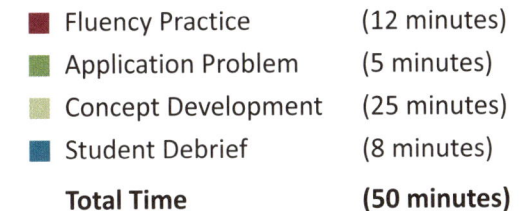

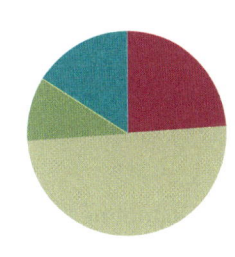

- ■ Fluency Practice   (12 minutes)
- ■ Application Problem   (5 minutes)
- ■ Concept Development   (25 minutes)
- ■ Student Debrief   (8 minutes)
-   **Total Time**   **(50 minutes)**

Fluency Practice (12 minutes)

- Dot Cards of 9 **K.CC.5, K.CC.2**   (4 minutes)
- Building Up to the Sprint Routine: Starting and Stopping at the Signal **K.CC.3**   (5 minutes)
- Show Me 1 More, 1 Less **K.CC.4c**   (3 minutes)

Dot Cards of 9 (4 minutes)

Materials: (T/S) Varied dot cards of 9 (Fluency Template)

T: (Show a card with 9 dots.) How many dots do you count? Wait for the signal to tell me. (Signal.)
S: 9.
T: How can you see them in two parts?
S: (Come forward to the card.) I saw 5 here and 4 here. → I saw 3 here and 6 here. → I saw 2 here and 7 here.

Repeat with other cards. Pass out the cards for students to work independently.

Building Up to the Sprint Routine: Starting and Stopping at the Signal (5 minutes)

Materials: (S) Lined writing paper

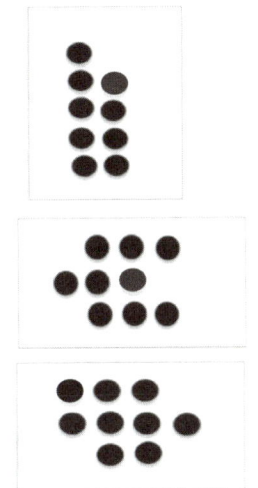

Note: Although the task is simple, this activity conditions students to stop working, even when they have not finished, and develops the self-regulation necessary for participating in math Sprints. Teaching the Sprint routine in stages may be time-consuming, but the investment is worthwhile.

Conduct as described in Lesson 16, but this time, increase the level of difficulty by having students write the numbers counting down from 10 to 0.

Lesson 19: Compare using *fewer than* and *the same as*.

Show Me 1 More, 1 Less (3 minutes)

Note: Students develop flexibility with the terms *more* and *less,* building upon the previous lesson and preparing for the current lesson.

- T: Show me three fingers the Math Way.
- S: (Hold up the left pinky, left ring finger, and left middle finger.)
- T: Now, show me 1 more.
- S: (Hold up the left pinky, left ring finger, left middle finger, and left index finger.)
- T: How many fingers are you showing me now?
- S: 4.
- T: We can say it like this, "3. 1 more is 4." Echo me, please.
- S: 3. 1 more is 4.
- T: New number. Show me 5.
- S: (Show open left hand.)
- T: Now, show me 1 less.
- S: (Hold up the left pinky, left ring finger, left middle finger, and left index finger.)
- T: How many fingers are you showing me now?
- S: 4.
- T: We can say it like this, "5. 1 less is 4." Echo me, please.
- S: 5. 1 less is 4.

Continue, and when students are ready, have them provide *1 more* and *1 less* statements on their own.

Application Problem (5 minutes)

Materials: (S) 1 small ball of clay

Use your clay to make six little pretend pancakes. How many people could you serve with your pancakes if you were going to have a tiny pancake party? What if another person joined them? Put your clay back together into a ball. Make new tiny pancakes so there are just enough. Talk about your cooking with your friend.

Note: This problem requires students to recalculate *just enough* in a change situation, providing an anticipatory set for the discussion of *less than* and *fewer than* in today's lesson.

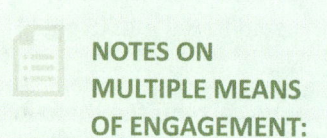

NOTES ON MULTIPLE MEANS OF ENGAGEMENT:

For students working below grade level, break the Application Problem down into smaller chunks, and ask students to practice the process. Show students what to do. Make a little pancake with the clay, and give it to a student while asking, "Is that enough? Did everyone get one? Do we need to make more?" Continue until students are able to complete this problem independently.

A STORY OF UNITS   Lesson 19   K•3

Concept Development (25 minutes)

Materials: (T) Box of markers (S) Bag of 5 pennies, bag of 10 loose linking cubes

- T: (Lay five markers on the table.) I am going to call up six students to be my helpers. (Say to each student, one at a time.) Please take a marker, and hold it up.
- S: (Sixth student.) I can't take one. There are none left!
- T: Oh, no! There are **fewer** markers **than** students! There are not enough markers so that each student can have one. (Hand sixth student a marker.) Now, are there enough?
- S: Yes! Everyone has one.
- T: Now, the number of markers is **the same as** the number of students. Each student has one. Please give me your markers, and return to your seats.

Repeat exercise several times, each time emphasizing the *fewer than* and *the same as* language, until all students have had a chance to participate.

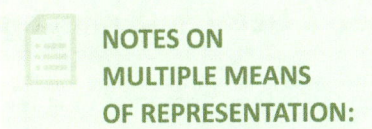

> **NOTES ON MULTIPLE MEANS OF REPRESENTATION:**
>
> Help English language learners by asking them questions that scaffold the concepts of *the same as* and *fewer than*. After the sixth student is left without a marker, ask, "Did everyone get a marker? Do we have a marker to give to everyone?" When the student receives a marker, ask, "Are there more students who need a marker? Are there markers left over?"

MP.6

- T: You have a bag of pennies and a bag of linking cubes. Please arrange the objects on your desk. What do you notice?
- S: We have more cubes. → There aren't as many pennies.
- T: How did you know? (Allow students to talk about their comparison strategies. Did they count them? Did they line them up to compare? Discuss all relevant strategies.)
- T: You are right! Echo me: "There are fewer pennies than cubes."
- S: There are fewer pennies than cubes.
- T: Put one cube back into your bag. Look at the cubes and pennies again. What do you notice?
- S: There are still more cubes!
- T: Echo me: "There are fewer pennies than cubes." (Repeat until there are five of each object on the desktops.)
- T: Look at your objects again. What do you notice?
- S: They are just the same!
- T: We have the same *number* of pennies as we do cubes! Echo me: "The *number* of pennies is the same as the *number* of cubes."
- S: The number of pennies is the same as the number of cubes.
- T: Please put your things away. We will do some more of these in our Problem Set now.

Lesson 19: Compare using *fewer than* and *the same as*.

Problem Set (10 minutes)

Students should do their personal best to complete the Problem Set within the allotted time.

Student Debrief (8 minutes)

Lesson Objective: Compare using *fewer than* and *the same as*.

The Student Debrief is intended to invite reflection and active processing of the total lesson experience.

Invite students to review their solutions for the Problem Set. They should check work by comparing answers with a partner before going over answers as a class. Look for misconceptions or misunderstandings that can be addressed in the Debrief. Guide students in a conversation to debrief the Problem Set and process the lesson.

Any combination of the questions below may be used to lead the discussion.

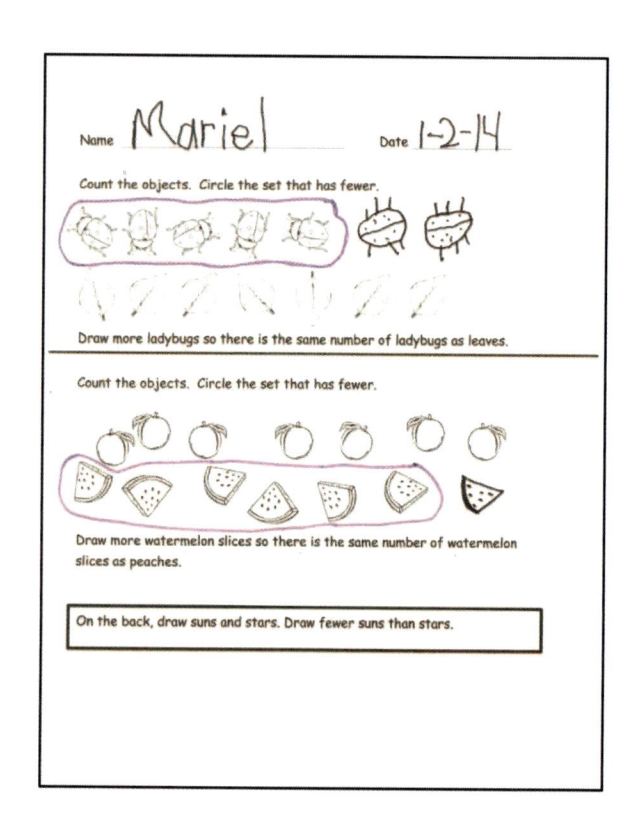

- How did you figure out how many pancakes to make when you had an extra guest?
- What strategies did you use to compare the cubes and the pennies on your desk?
- How did you know when there were **the same** number of cubes **as** pennies?
- In the Problem Set, how did you know which set had **fewer than** the other? How did you draw to make the same number of ladybugs as leaves?
- How many suns and stars did you draw on the back of your Problem Set? Were there fewer suns or stars?
- Talk to your neighbor about how your drawings were different. Did your partner have more suns or stars? Did you have more suns or stars? Count all of your suns and stars. How many did you have? Check with your partner. Who had fewer than his or her partner? Did anyone have the same number as his or her partner?
- What important math vocabulary did we use to communicate precisely?

Name _____ Date _____

Count the objects. Circle the set that has fewer.

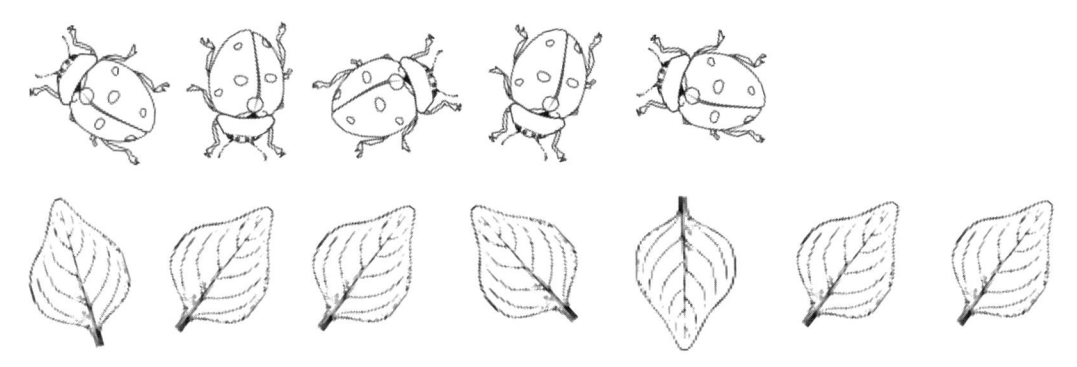

Draw more ladybugs so there are the same number of ladybugs as leaves.

___

Count the objects. Circle the set that has fewer.

Draw more watermelon slices so there are the same number of watermelon slices as peaches.

On the back, draw suns and stars. Draw fewer suns than stars.

Name _____   Date _____

Draw another bird so there are the same number of birds as bird cages.

On the back of your paper, draw 5 dogs.

Draw dog houses so there are *fewer* dog houses than dogs.

Draw bones so there are the *same* number of bones as dogs.

Lesson 19:   Compare using *fewer than* and *the same as*.

A STORY OF UNITS

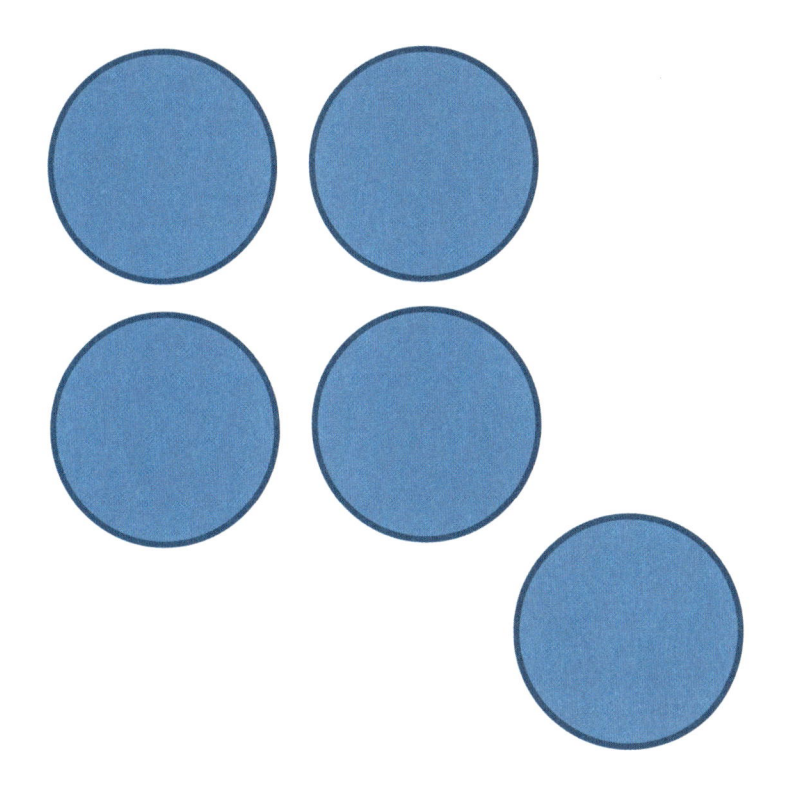

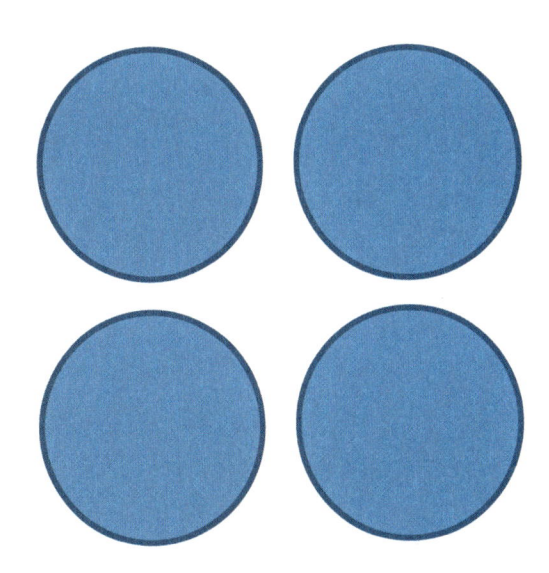

Lesson 19 Fluency Template K•3

dot cards of 9

Lesson 19: Compare using *fewer than* and *the same as*.

199

dot cards of 9

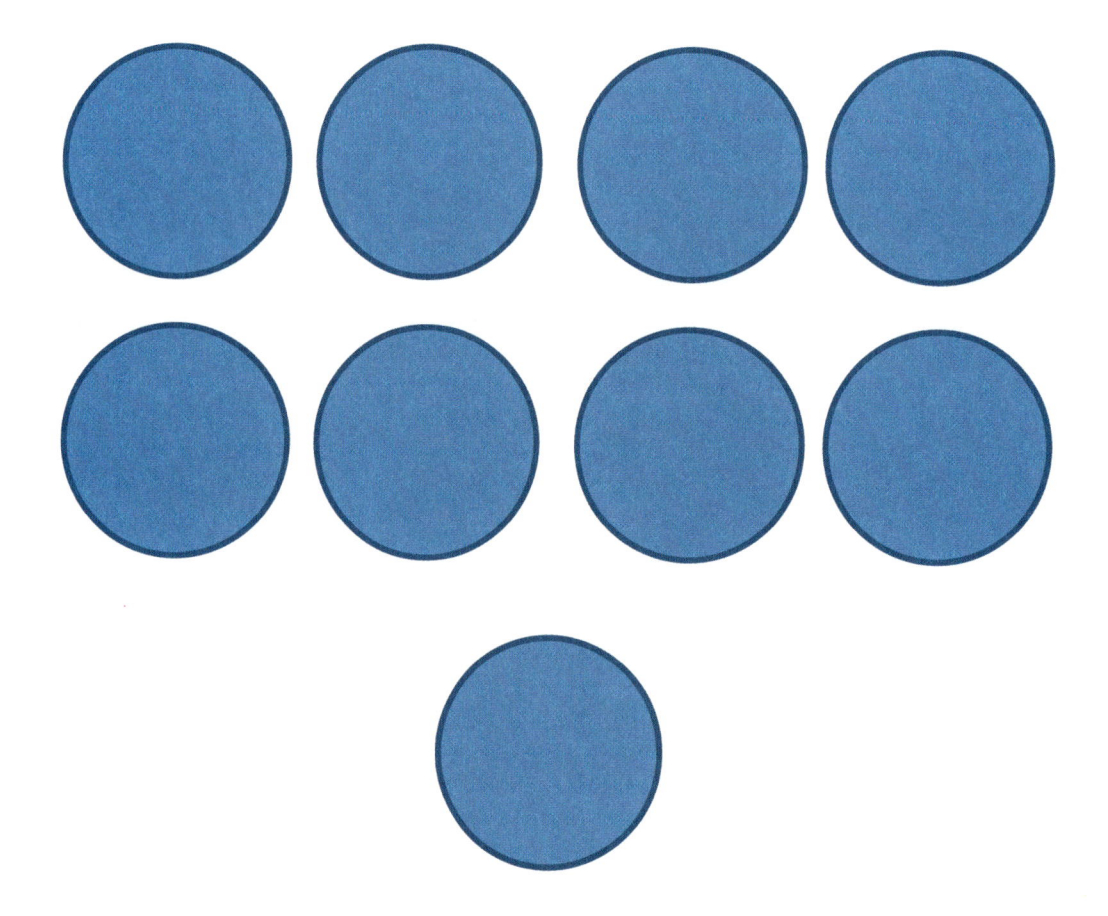

A STORY OF UNITS Lesson 19 Fluency Template K•3

dot cards of 9

Lesson 19: Compare using *fewer than* and *the same as*.

201

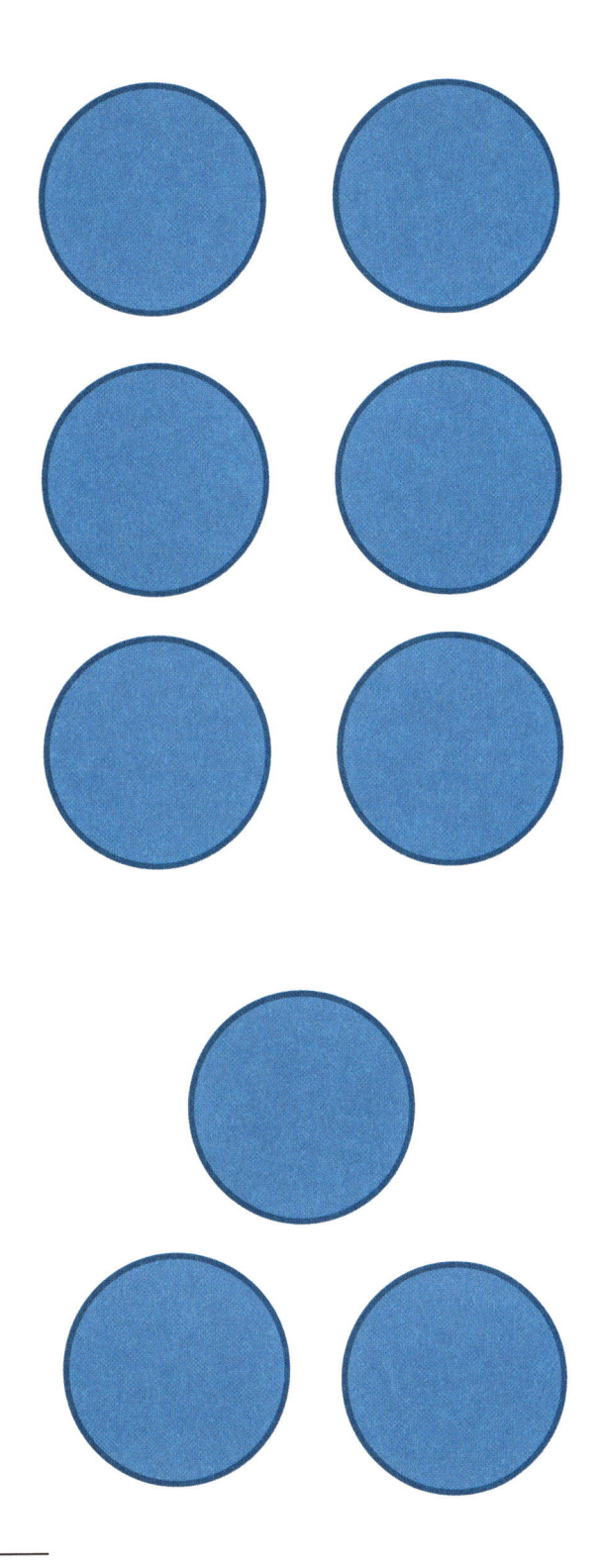

A STORY OF UNITS

Lesson 19 Fluency Template K•3

dot cards of 9

Lesson 19: Compare using *fewer than* and *the same as*.

EUREKA MATH

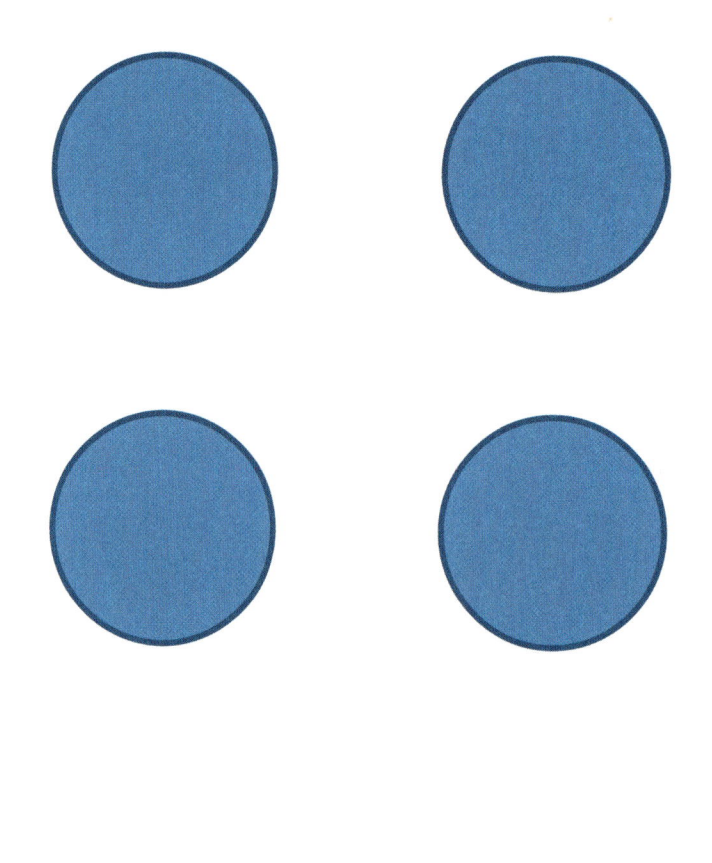

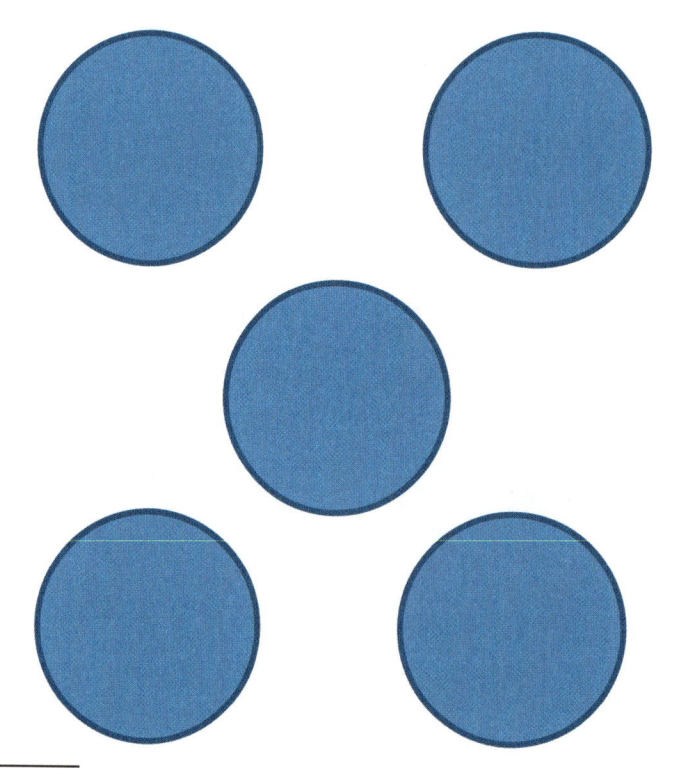

dot cards of 9

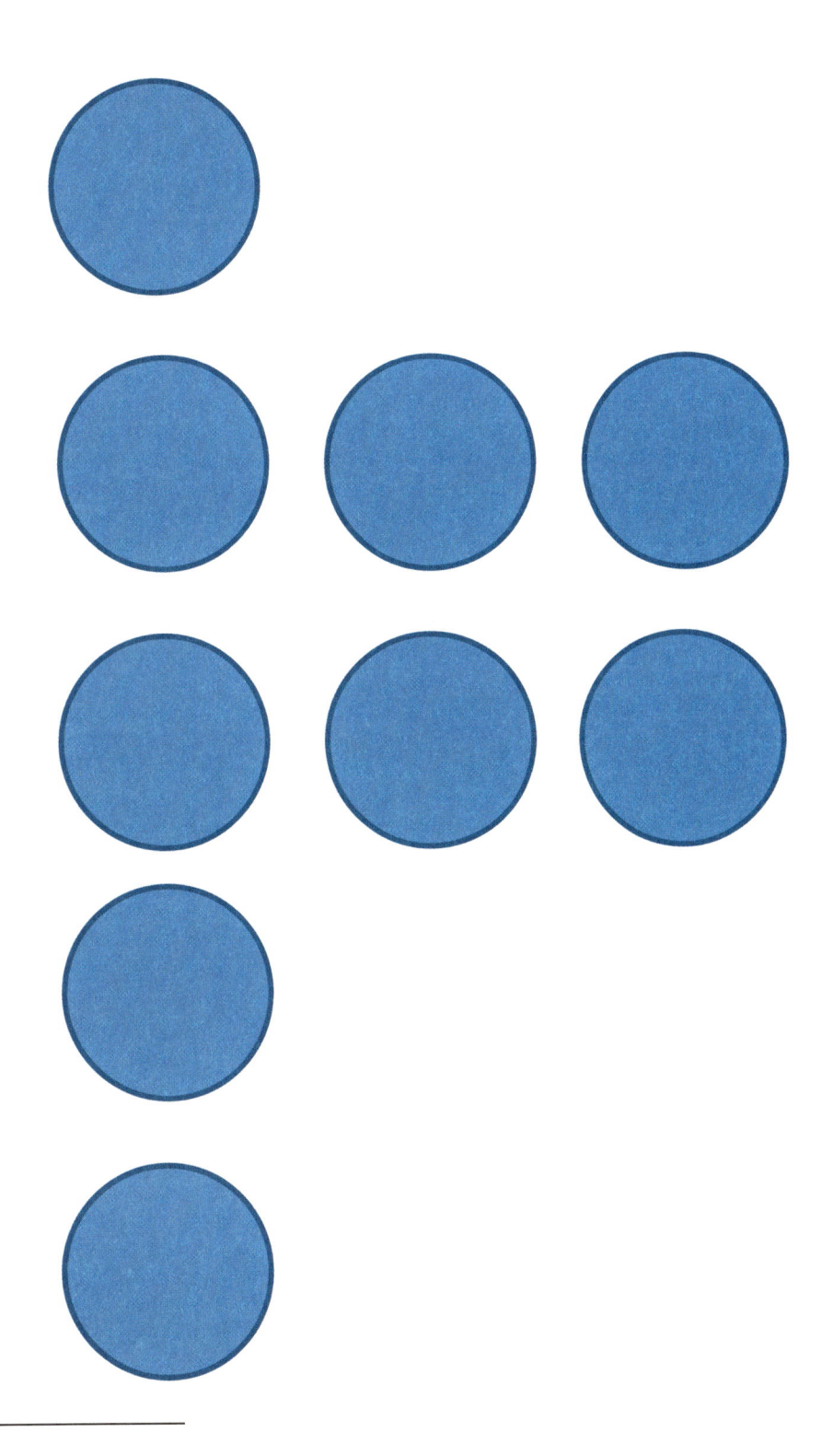

dot cards of 9

Lesson 19: Compare using *fewer than* and *the same as*.

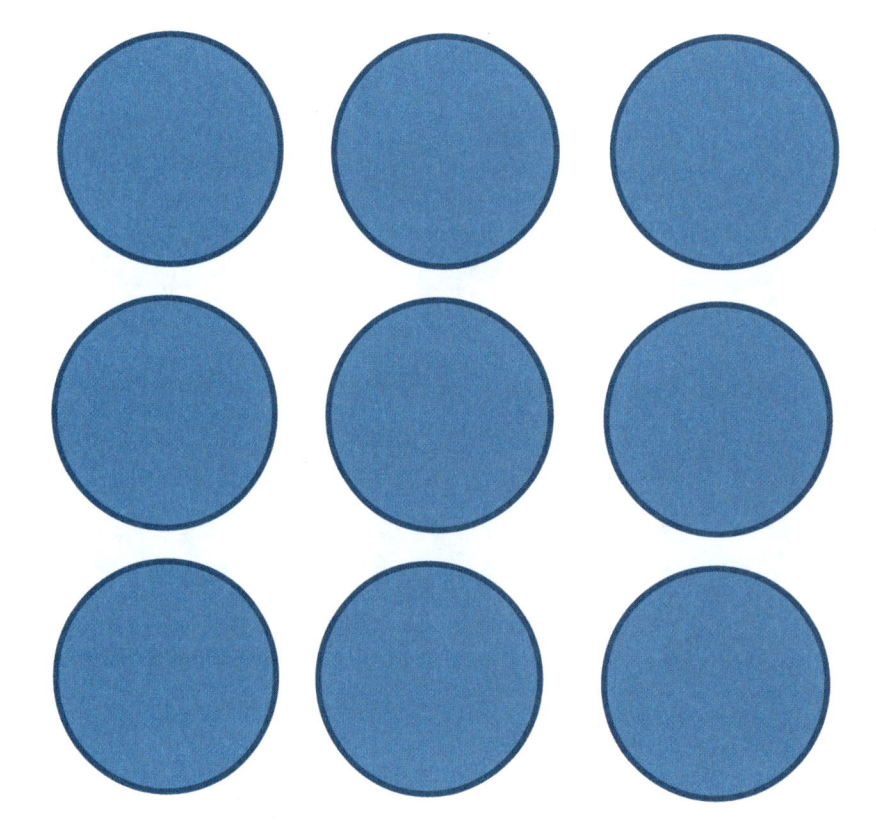

dot cards of 9

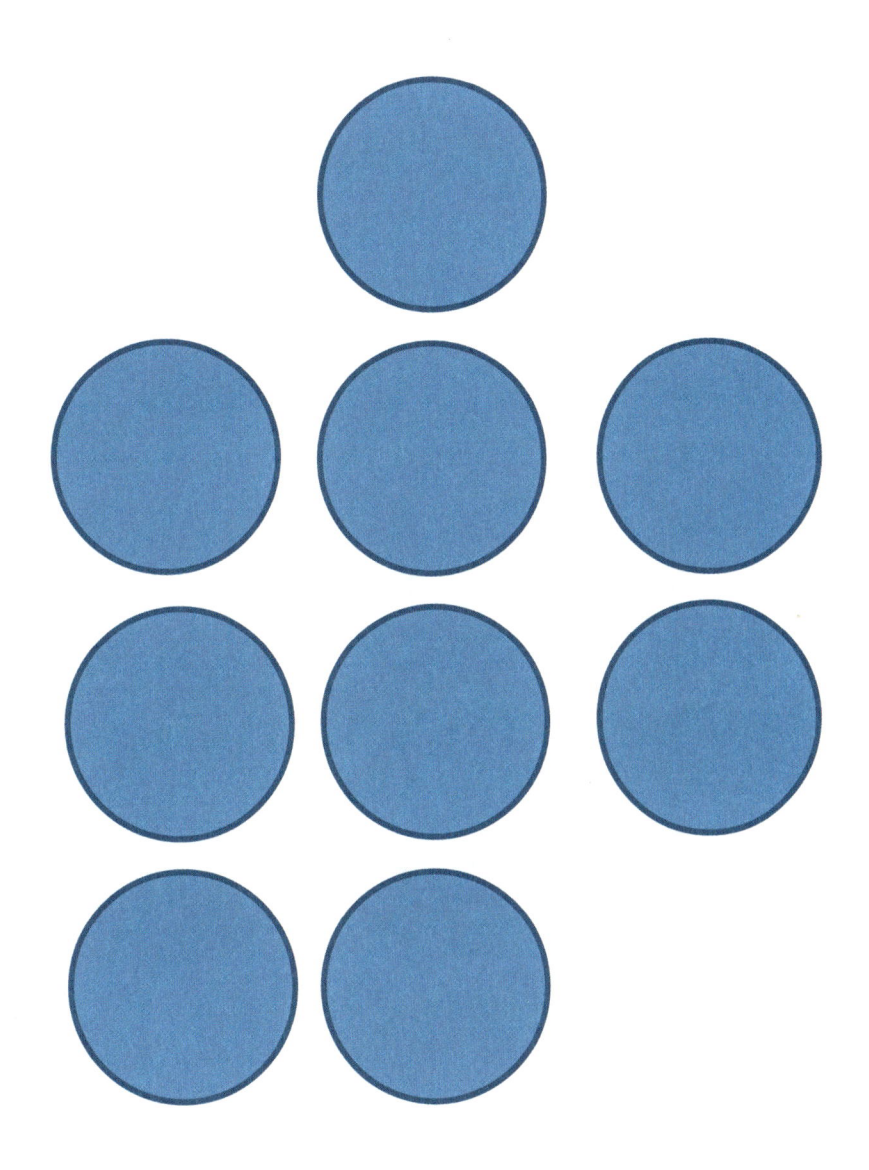

dot cards of 9

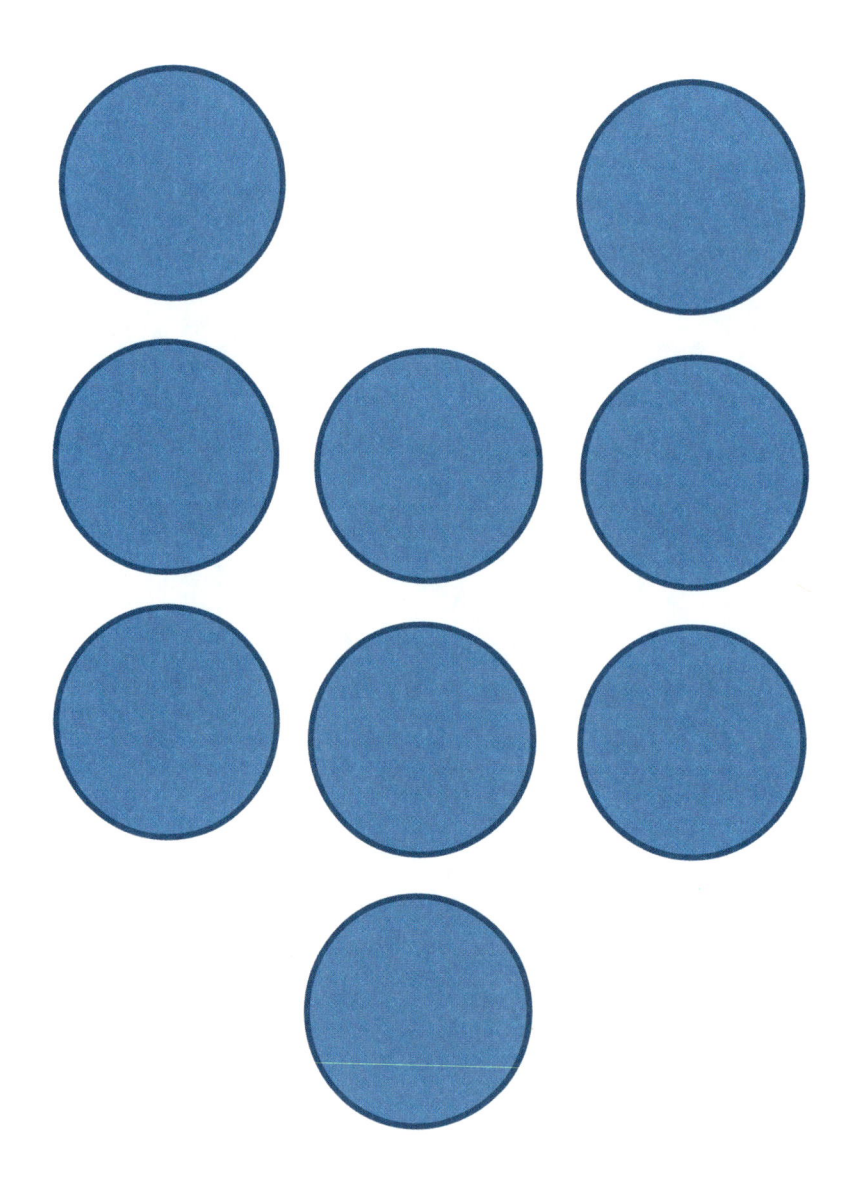

A STORY OF UNITS

Mathematics Curriculum

GRADE K • MODULE 3

Topic F
Comparison of Sets Within 10

K.CC.6, K.CC.7, K.CC.4c, K.MD.2

Focus Standards:	K.CC.6	Identify whether the number of objects in one group is greater than, less than, or equal to the number of objects in another group, e.g., by using matching and counting strategies. (Include groups with up to ten objects.)
	K.CC.7	Compare two numbers between 1 and 10 presented as written numerals.
Instructional Days:	5	
Coherence	-Links from: GPK–M4	Comparison of Length, Weight, Capacity, and Numbers to 5
	-Links to: G1–M3	Ordering and Comparing Length Measurements as Numbers

Topic F opens with students shifting from comparison of lengths to comparison of numbers. As students build their confidence by directly comparing the lengths of a pencil and a crayon, they increase their readiness in later grades to indirectly compare length units. "The pencil is longer than the crayon because 7 cubes are more than 4 cubes."

In Lesson 20, students relate *more* and *less* to length: "A stick of 7 cubes is longer than a stick of 3 cubes; 7 is more than 3. A stick of 3 cubes is shorter than a stick of 7 cubes; 3 is less than 7."

In Lesson 21, students take two sticks, break them into cubes, and compare the sets. "Which set has more objects? This set has more than that set."

In Lessons 22–24, students create and identify sets that have the same number of objects, 1 more object, and 1 fewer object.

A Teaching Sequence Toward Mastery of Comparison of Sets Within 10

Objective 1: Relate *more* and *less* to length.
(Lesson 20)

Objective 2: Compare sets informally using *more*, *less*, and *fewer*.
(Lesson 21)

Objective 3: Identify and create a set that has the same number of objects.
(Lesson 22)

Objective 4: Reason to identify and make a set that has 1 more.
(Lesson 23)

Objective 5: Reason to identify and make a set that has 1 less.
(Lesson 24)

| A STORY OF UNITS | Lesson 20 K•3 |

Lesson 20

Objective: Relate *more* and *less* to length.

Suggested Lesson Structure

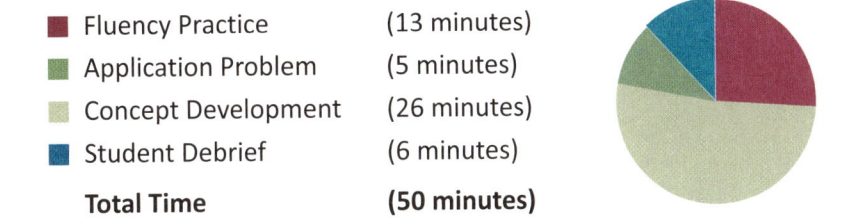

- ■ Fluency Practice   (13 minutes)
- ■ Application Problem   (5 minutes)
- ■ Concept Development   (26 minutes)
- ■ Student Debrief   (6 minutes)
- **Total Time**   **(50 minutes)**

Fluency Practice (13 minutes)

- Building up to the Sprint Routine: Observing and Noticing **K.CC.5**   (8 minutes)
- Building *1 More* and *1 Less* Trains **K.CC.4c**   (5 minutes)

Building Up to the Sprint Routine: Observing and Noticing (8 minutes)

Materials: (T) Count and Circle How Many Sprint (project for students to view), framed portrait of the teacher at 5–6 years old

Note: Teaching the Sprint routine in stages may be time-consuming, but the investment is worthwhile. Providing students this opportunity to observe and reflect increases motivation, enthusiasm, and success in this strong fluency exercise. Students complete their first Sprint in Lesson 21.

1. Tell students to watch the teacher do a math race called a Sprint as if the teacher were a student back in Kindergarten. Place the portrait on the desk where the teacher is working to remind students of the role. If possible, have an assistant play the role of the teacher delivering the Sprint.
2. At the start signal, turn the paper over, and begin working. Start at the top left corner with the hearts, and continue working down the hearts column. At the bottom of the hearts column, start again at the top of the stars column.
3. At the signal, stop and hold the pencil up, just as students have practiced in previous Sprint preparation exercises. Be careful to display a positive demeanor even though the task is not finished. Possibly pretend to wipe away sweat from the brow to emphasize working with intensity, and smile with satisfaction for having made such a strong effort! (Be sure to ask the assistant playing the role of the teacher to limit the timeframe, or set a timer, so that the teacher comes very close to completing the Sprint but does not quite finish.)
4. While reviewing the answers (now projected on the board), students circle correct answers in the air with their fingers, along with the teacher, energetically shouting "Yes!" for each correct answer. The entire class counts the number of correct problems chorally and writes the number in the air as the teacher writes it at the top of the page.

A STORY OF UNITS

Lesson 20 K•3

5. Conclude the observation and role play. Then, gather the group at the rug to debrief the process. The following are suggested questions to guide the conversation:
   - When did the teacher (playing the role of a Kindergarten student) begin working on the problems?
   - Which problems did the teacher do first—the hearts or stars? (This question helps students realize that the Sprint is designed to be completed working down the columns, not across the rows.)
   - What did the teacher do when the timer sounded (or other stopping signal occurred)?
   - How did the teacher react at the end? (Emphasize that the goal is maximum effort and efficiency, not completion. Begin setting expectations for social and emotional behaviors during Sprints.)

Optional: Create a few intentional errors. Let students know to expect this beforehand. Tell them to be ready to explain what went wrong, being careful to avoid having students perceive the teacher as acting foolishly.

Building *1 More* and *1 Less* Trains (5 minutes)

Note: In this activity, students connect increasing and decreasing length to increasing and decreasing numerical value.

Conduct the activity as described in Lesson 15, but now, have students build and disassemble the cubes horizontally, similar to a train.

Application Problem (5 minutes)

Materials: (S) Square path letter trains (Template)

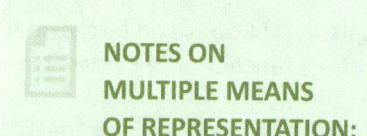

> **NOTES ON MULTIPLE MEANS OF REPRESENTATION:**
>
> Model the directions of the Application Problem for English language learners. Show them one step at a time what to do, saying, "Start at the box above the star," while pointing to the star, etc.

- Write your first name in the top set of boxes, one letter in each box. Start at the box above the star.
- Write your last name in the bottom set of boxes, one letter in each box. Start at the box above the star.
- Which of your trains has more letter passengers? Which passenger train is longer?
- Which of your trains has fewer passengers? Which passenger train is shorter?
- Talk about your trains with your partner. Are your partner's trains similar to yours?
- Did anyone's train not have enough room for all of the letter passengers?

Note: The comparison of the lengths of the letter trains serves as the anticipatory set for the concrete work in today's lesson.

Lesson 20: Relate *more* and *less* to length.

EUREKA MATH

Concept Development (26 minutes)

Materials: (T/S) Bag of 20 linking cubes, 10-sided die

T: I am going to make a stick of 7 linking cubes. Student A, could you please make a stick of 3 linking cubes?

T: Which one of our sticks is longer?

S: Your 7-stick!

T: Yes! (Demonstrate.) The 7-stick is longer than the 3-stick, and the 3-stick is shorter than the 7-stick. How did you know? (Discuss comparison strategies. Did they line them up in their minds? Did they mentally match one-to-one? Did they estimate?) Let's count the cubes on each side. (Count chorally, and write the numbers on the board.) What do you notice about the numbers 7 and 3? Which is more?

S: 7 is more! 3 is less than 7.

T: 7 is more than 3. 3 is less than 7. How can you be sure?

S: I can see that 7 is longer.

T: You are right! A 7-stick is longer than a 3-stick. (You may wish to match the sets of cubes one-to-one to demonstrate the validity of their argument, showing that there are still some left after pairs have been removed.)

T: Now, I'm going to make a 5-stick. Student C is going to make an 8-stick. Let's hold our sticks up. Which stick is longer? Which is shorter? Which stick has more? Which has less? How did you know? (Allow time for discussion.)

MP.2

T: We are going to play a game. Roll the die with your partner. Make a stick using the same number of cubes as the dots that your die shows. Roll the die again, and make another stick with that number of cubes. Compare the length of your sticks. Which is longer? Finally, take your sticks apart. Put the sets of cubes on the table, and compare them. Which set has more?

T: Count each set of cubes, and write the number on a small card. Compare the numbers. Which is more? Which is less? (Circulate during the activity to encourage correct mathematical vocabulary and to ensure accuracy of numerical representations.)

T: Roll the die again, and make two new sticks to compare! (Repeat as long as time allows.)

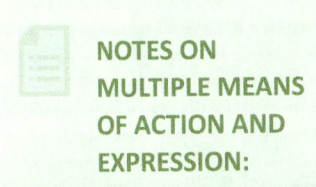

NOTES ON MULTIPLE MEANS OF ACTION AND EXPRESSION:

Scaffold the activity for struggling students by modeling how to play the game. Play one round with a student or group of students until they are clear about what they need to do. Watch them play one round to ensure that they are on the correct track.

Problem Set (10 minutes)

Students should do their personal best to complete the Problem Set within the allotted time.

Student Debrief (6 minutes)

Lesson Objective: Relate *more* and *less* to length.

The Student Debrief is intended to invite reflection and active processing of the total lesson experience.

Invite students to review their solutions for the Problem Set. They should check work by comparing answers with a partner before going over answers as a class. Look for misconceptions or misunderstandings that can be addressed in the Debrief. Guide students in a conversation to debrief the Problem Set and process the lesson.

Any combination of the questions below may be used to lead the discussion.

- What are some of the ways you could tell which set had more cubes in our activity?
- If one stick has more cubes than another, will it be longer than the other?
- How can you compare the number of cubes in one set to another set? How can you tell which number is more?
- Talk to your partner about the chain you made by rolling the die for your Problem Set. What numbers did you roll? How did you know which had fewer beads?
- For the back of the Problem Set, what numbers did you roll? What did you do to make sure you drew the same number of beads as the number you rolled?
- If one stick has fewer cubes than another, will it be heavier or lighter than the other?

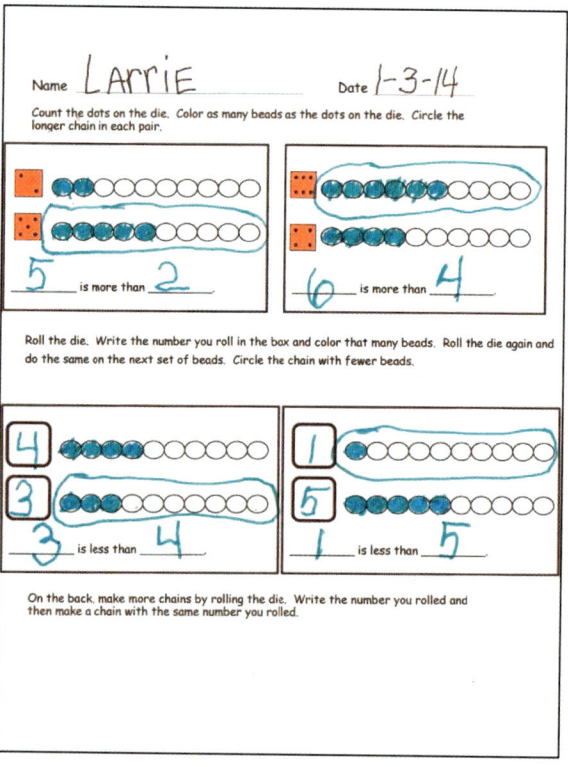

Count and Circle How Many

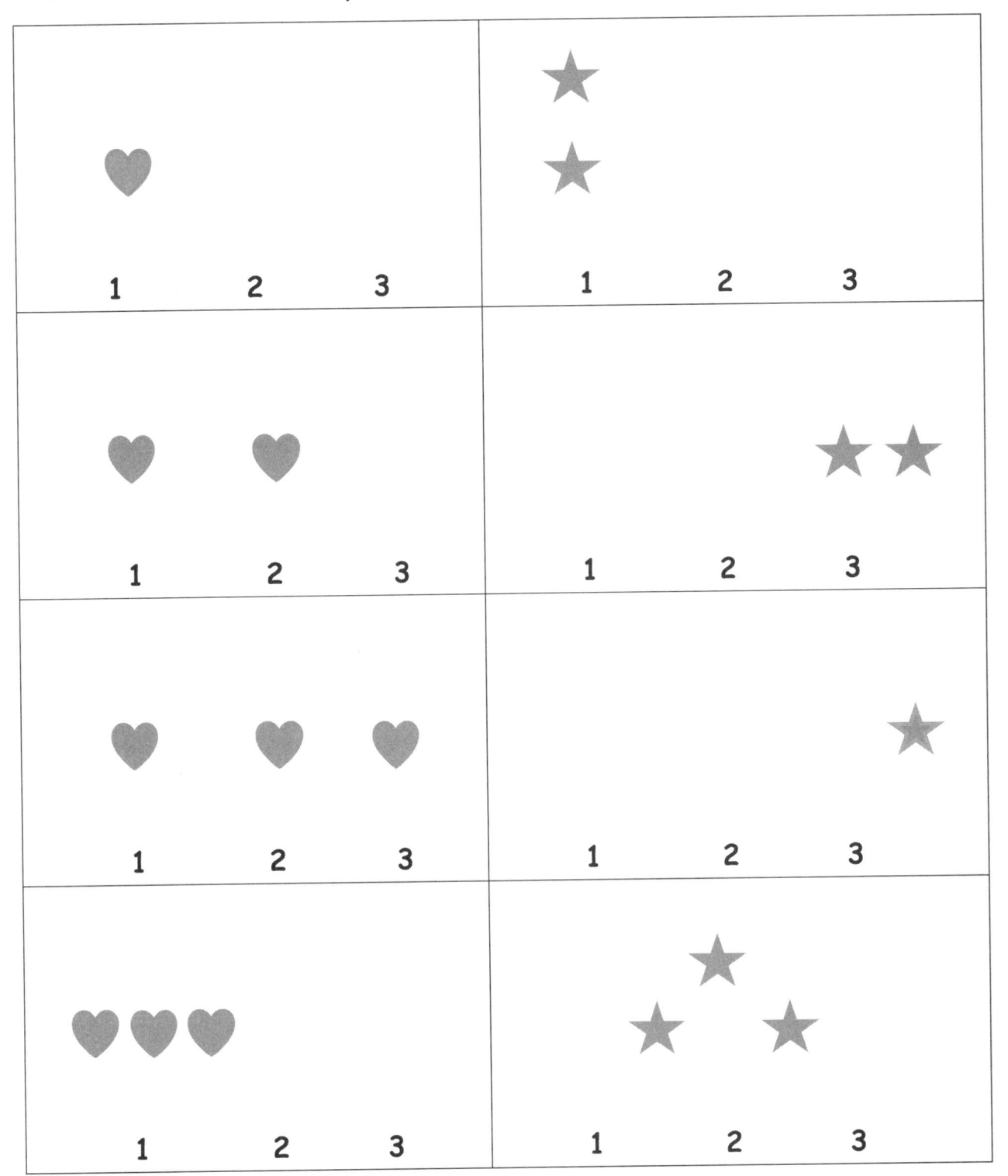

♥	★ ★
1 2 3	1 2 3
♥ ♥	★ ★
1 2 3	1 2 3
♥ ♥ ♥	★
1 2 3	1 2 3
♥♥♥	★ ★ ★
1 2 3	1 2 3

Lesson 20: Relate *more* and *less* to length.

A STORY OF UNITS

Lesson 20 Problem Set K•3

Name _____   Date _____

Count the dots on the die. Color as many beads as the dots on the die. Circle the longer chain in each pair.

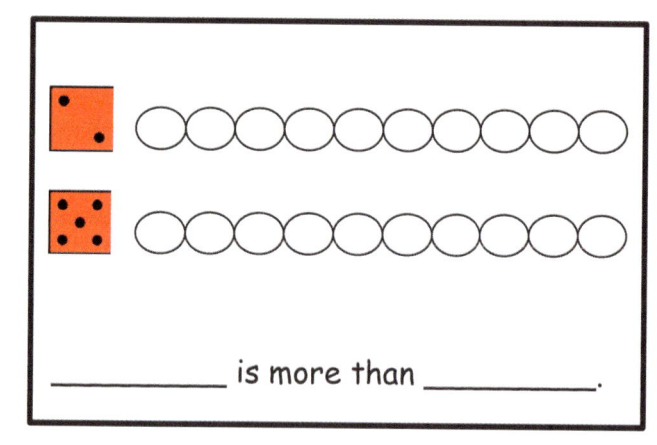

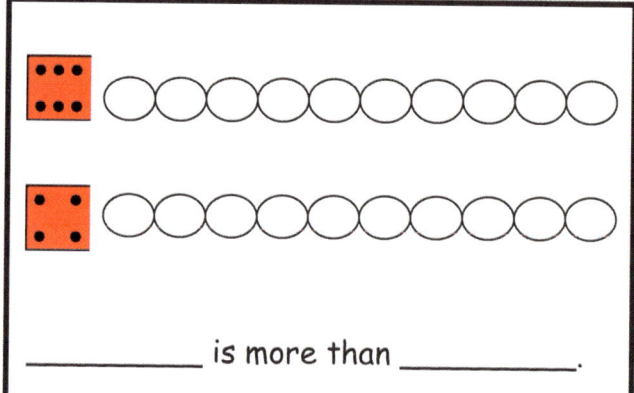

_____ is more than _____.

_____ is more than _____.

Roll the die. Write the number you roll in the box, and color that many beads. Roll the die again, and do the same on the next set of beads. Circle the chain with fewer beads.

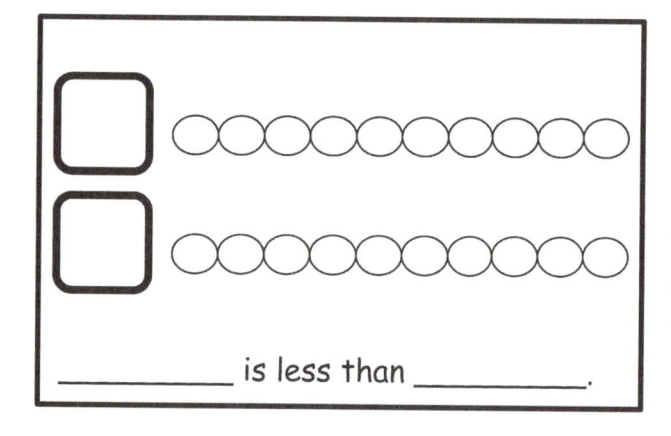

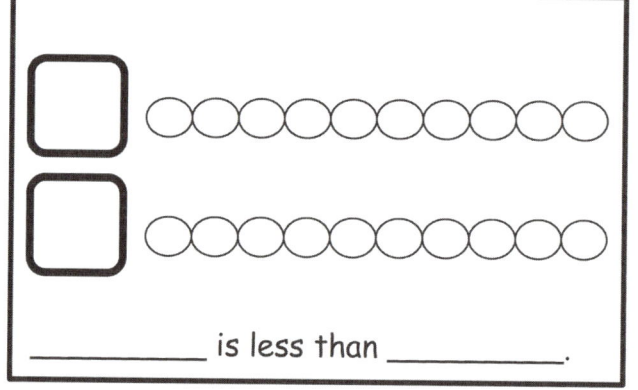

_____ is less than _____.

_____ is less than _____.

On the back, make more chains by rolling the die. Write the number you rolled, and then make a chain with the same number you rolled.

214 Lesson 20: Relate *more* and *less* to length.

EUREKA MATH

A STORY OF UNITS   Lesson 20 Homework K•3

Name _____   Date _____

On the first chain, color the first 3 beads blue.
On the next chain, color more than 3 beads red.
How many beads did you color red? Write the number in the box.

_____ red beads is more than 3.

On the first chain, color the first 5 beads green.
On the next chain, color fewer than 5 beads yellow.
How many beads did you color yellow? Write the number in the box.

_____ yellow beads is fewer than 5.

Color 2 beads brown in the first column.

Color more than 2 beads blue in the second column.

How many beads did you color in the second column? Write the number in the box.

_____ blue beads is more than 2.

Lesson 20:   Relate *more* and *less* to length.   215

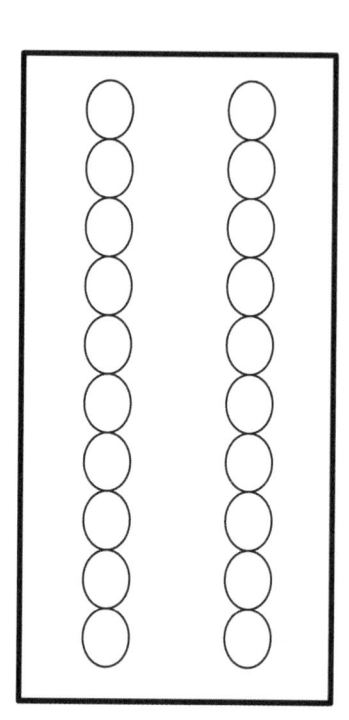

A STORY OF UNITS   Lesson 20 Homework K•3

Color 9 beads red in the first column.

Color fewer than 9 beads green in the second column.

How many beads did you color in the second column? Write the number in the box.

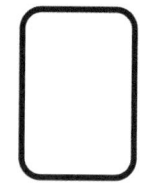

_____ green beads is fewer than 9.

Draw a chain with more than 3 beads but fewer than 10 beads.

Draw a chain that has fewer than 10 beads but more than 4 beads.

Lesson 20: Relate *more* and *less* to length.

A STORY OF UNITS

Lesson 20 Template K•3

square path letter trains

Lesson 20: Relate *more* and *less* to length.

217

Lesson 21

Objective: Compare sets informally using *more, less*, and *fewer*.

Suggested Lesson Structure

- ■ Fluency Practice   (13 minutes)
- ■ Application Problem   (5 minutes)
- ■ Concept Development   (25 minutes)
- ■ Student Debrief   (7 minutes)
- **Total Time**   **(50 minutes)**

Fluency Practice (13 minutes)

- My First Sprint **K.CC.5**   (8 minutes)
- Finger Number Pairs **K.CC.4a**   (5 minutes)

My First Sprint (8 minutes)

Materials: (S) 1 copy of the Count and Circle How Many Sprint (Lesson 20)

Note: This activity allows students to become comfortable with Sprint procedures as they work on this simple task with confidence.

- T: Today, you will get to do a math race called a Sprint. (Remind students of the previous day's activity.) Take out your pencil and one crayon of any color.
- T: (Distribute the Sprint papers face down.) On your mark, get set, go!
- T: (Ring the bell, or give another signal for students to stop. Although it will not be necessary to time the students in this short practice Sprint, be sure to give the stop signal before students finish so as to not develop the expectation of finishing every time.) Pencils up!
- T: Pencils down and crayons up! It's time to check answers. What do you do if the answer is right?
- S: Circle it.
- T: What do you say?
- S: Yes!
- T: We'll begin with the hearts. Ready? 1.
- S: Yes!
- T: 2.
- S: Yes!

A STORY OF UNITS   Lesson 21   K•3

Continue checking the remaining answers. Then, have students count the number correct and write the number at the top. Maintain the celebratory mood. Praise students for learning a new procedure, as well as their strong effort and hard work. Note that only one Sprint is delivered this time. The two-part Sprint is introduced in a future lesson.

Troubleshooting: If students work across instead of down the columns, create a green arrow down the left-hand side and a red arrow along the right-hand side to indicate where to start and stop. If students have difficulty circling the answers quickly, give them a highlighter, and allow them to swipe the correct answer.

Finger Number Pairs (5 minutes)

Note: This activity ensures that students do not become overly reliant on counting the Math Way and gives them yet another method of breaking apart numbers, essential to the work of the next module.

Conduct as outlined in Lesson 18, but this time, invite students to explain why certain combinations cannot be shown on two hands. A student might say, "I can show 10 as 5 on one hand and 5 on the other, but I can't show 10 as 6 and 4." Guide them to use some of their newly acquired vocabulary and be precise with respect to explaining their thoughts.

Application Problem (5 minutes)

Materials: (S) Linking cubes, dry erase marker

Use your dry erase markers to write the letters of your name on linking cubes. Make a train out of your cubes. Compare your train to at least one friend's train. Which train is longer? Count the cubes in your trains. Which number is more? Which number is less?

Note: This extension of yesterday's Application Problem serves as an introductory informal set comparison for today's lesson. When comparing a number of discrete objects, use the word *fewer*. When comparing numbers, use the word *less*.

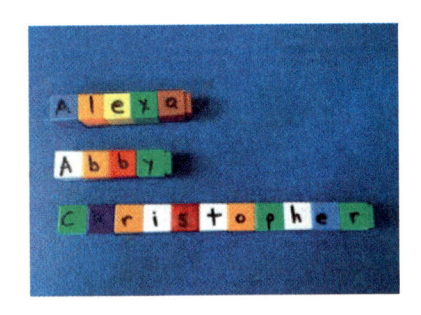

Concept Development (25 minutes)

Materials: (T) Shapes (Template 1) cut out and arranged in rows on the board (S) More than, fewer than recording sheet (Template 2)

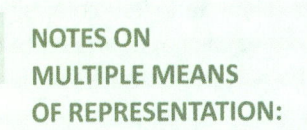

9 ▪ square
6 ● circle
4 ⬢ hexagon
6 ▲ triangle
7 ▬ rectangle

NOTES ON MULTIPLE MEANS OF REPRESENTATION:

English language learners benefit from seeing the names of the shapes as the teacher introduces and discusses them for the lesson. For each set of shapes, include *square, circle, triangle, hexagon,* and *rectangle*. Students can focus on *how many* of each shape is present rather than focusing on trying to produce their names.

Lesson 21:   Compare sets informally using *more, less,* and *fewer*.   219

EUREKA MATH

© 2015 Great Minds. eureka-math.org
GK-M3-TE-B3-1.3.1-01.2016

| | A STORY OF UNITS | Lesson 21 | K•3 |

Note: While the importance of definitions is not necessarily stressed in Kindergarten (recognition is intuitive at this stage), a square is still treated as a special type of rectangle. If asked how many rectangles, students might initially respond by saying 7, when, in actuality, there are 16.

MP.2

T: What do you notice on the board today?
S: I see shapes! → There are all different kinds.
T: What types of shapes do you see on the board? (Use this as an opportunity to discuss and review the shape types from Kindergarten Module 2.)
T: Are there more squares or triangles?
S: There are **more** squares **than** triangles.
T: How do you know?
S: The squares look bigger. → I counted them. (Discuss relevant strategies.)
T: Are there fewer circles or hexagons? (Continue informally comparing sets of shapes, and encourage students to discuss their strategies for finding more or less than.)
T: Which two groups have the same number of shapes?
S: The circles and triangles! → There are six circles and six triangles.
T: Let's compare our sets of shapes on the recording sheet. In each row, count how many of the shapes are on the board. Then, draw a shape that makes each sentence true. (Demonstrate. Pass out recording sheets, and circulate to ensure accuracy in terms of counting and comparison.)

Problem Set (10 minutes)

Students should do their personal best to complete the Problem Set within the allotted time.

Note: Give students step-by-step directions while completing the Problem Set. First, color all of the shapes. Then, count *how many* of each shape, and write the number in the box. Finally, use the first page of the Problem Set to complete the second page.

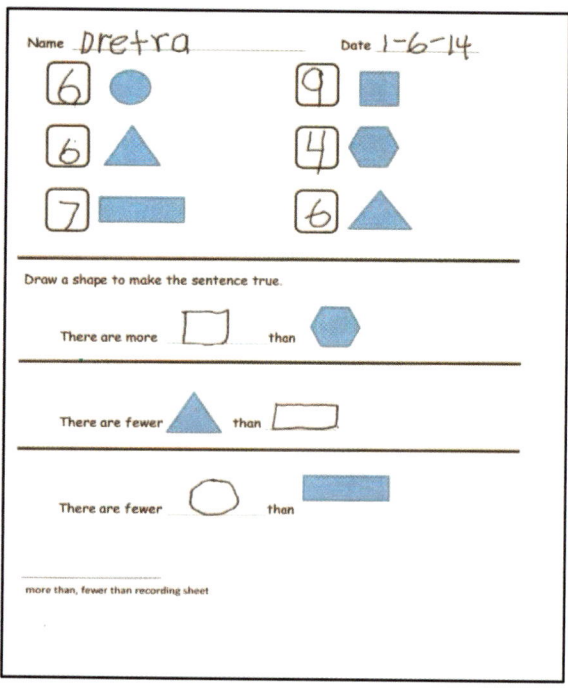

NOTES ON MULTIPLE MEANS FOR ACTION AND EXPRESSION:

For students working above grade level, expand the lesson by asking them to arrange the groups of shapes from least to greatest and explain how they knew which set had the least and which set had the most. Ask students to draw more circles so that there are the same numbers of circles as squares, etc.

220   Lesson 21:   Compare sets informally using *more*, *less*, and *fewer*.

A STORY OF UNITS   Lesson 21   K•3

Student Debrief (7 minutes)

Lesson Objective: Compare sets informally using *more*, *less*, and *fewer*.

The Student Debrief is intended to invite reflection and active processing of the total lesson experience.

Invite students to review their solutions for the Problem Set. They should check work by comparing answers with a partner before going over answers as a class. Look for misconceptions or misunderstandings that can be addressed in the Debrief. Guide students in a conversation to debrief the Problem Set and process the lesson.

Any combination of the questions below may be used to lead the discussion.

- Were there **more** circles **than** hexagons? Were there more squares than triangles?
- Were there **fewer** hexagons **than** triangles? Were there fewer rectangles than triangles?
- Which sets of shapes on the board had the same number?
- On the Problem Set, were there more circles than triangles? Were there fewer hexagons than rectangles?
- What new (or significant) math vocabulary did we use today to communicate precisely?

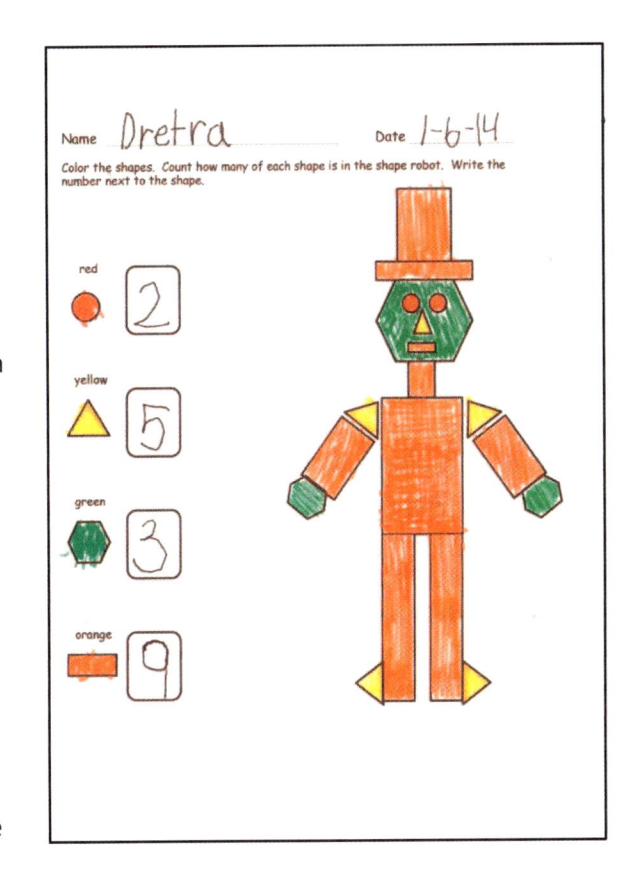

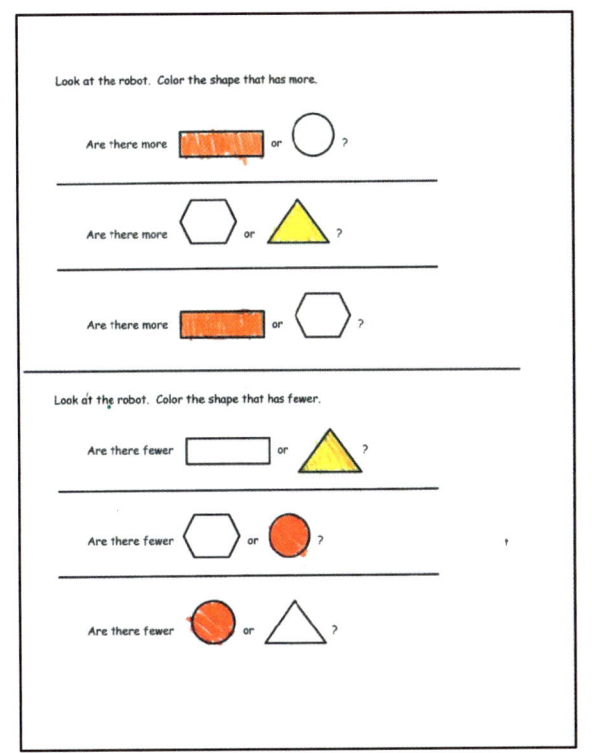

Lesson 21: Compare sets informally using *more*, *less*, and *fewer*.   221

A STORY OF UNITS

Lesson 21 Problem Set   K•3

Name _____   Date _____

Color the shapes. Count how many of each shape is in the shape robot. Write the number next to the shape.

Red ○ ☐

Yellow △ ☐

Green ⬡ ☐

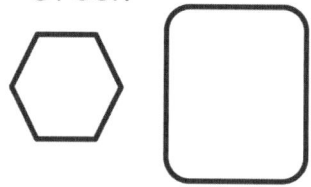

Orange ▭ ☐

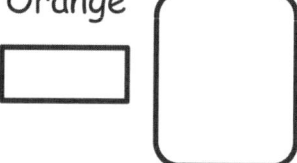

222   Lesson 21:   Compare sets informally using *more, less,* and *fewer.*

EUREKA MATH

A STORY OF UNITS

Lesson 21 Problem Set K•3

Look at the robot. Color the shape that has more.

Are there more ▭ or ◯ ?

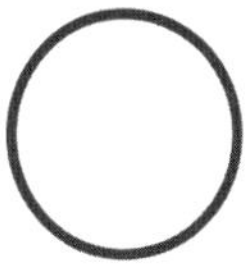

Are there more ⬡ or △ ?

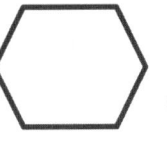

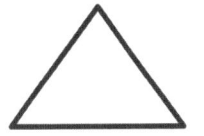

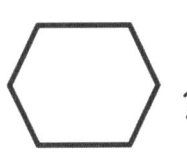

Are there more ▭ or ⬡ ?

Look at the robot. Color the shape that has fewer.

Are there fewer ▭ or △ ?

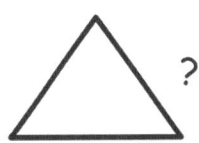

Are there fewer ⬡ or ◯ ?

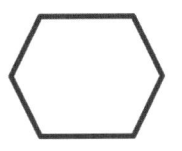

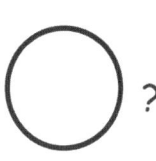

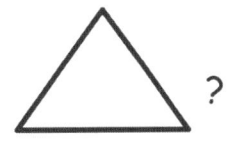

Are there fewer ◯ or △ ?

Lesson 21: Compare sets informally using *more*, *less*, and *fewer*.

223

A STORY OF UNITS　　　　　　　　　　　　　　　　　Lesson 21 Homework　K•3

Name _____   Date _____

Which has more? The 🛒 or 🚲 ?

Circle the set that has more.

Which has fewer? The  🐧 or 🛹 ?

Circle the set that has fewer.

Which has fewer? The  ★ or ☾ ?

Circle the set that has fewer.

On the back of your paper, draw a set of 5 books. Draw some apples. Are there fewer apples or fewer books?

Lesson 21: Compare sets informally using *more*, *less*, and *fewer*.

A STORY OF UNITS

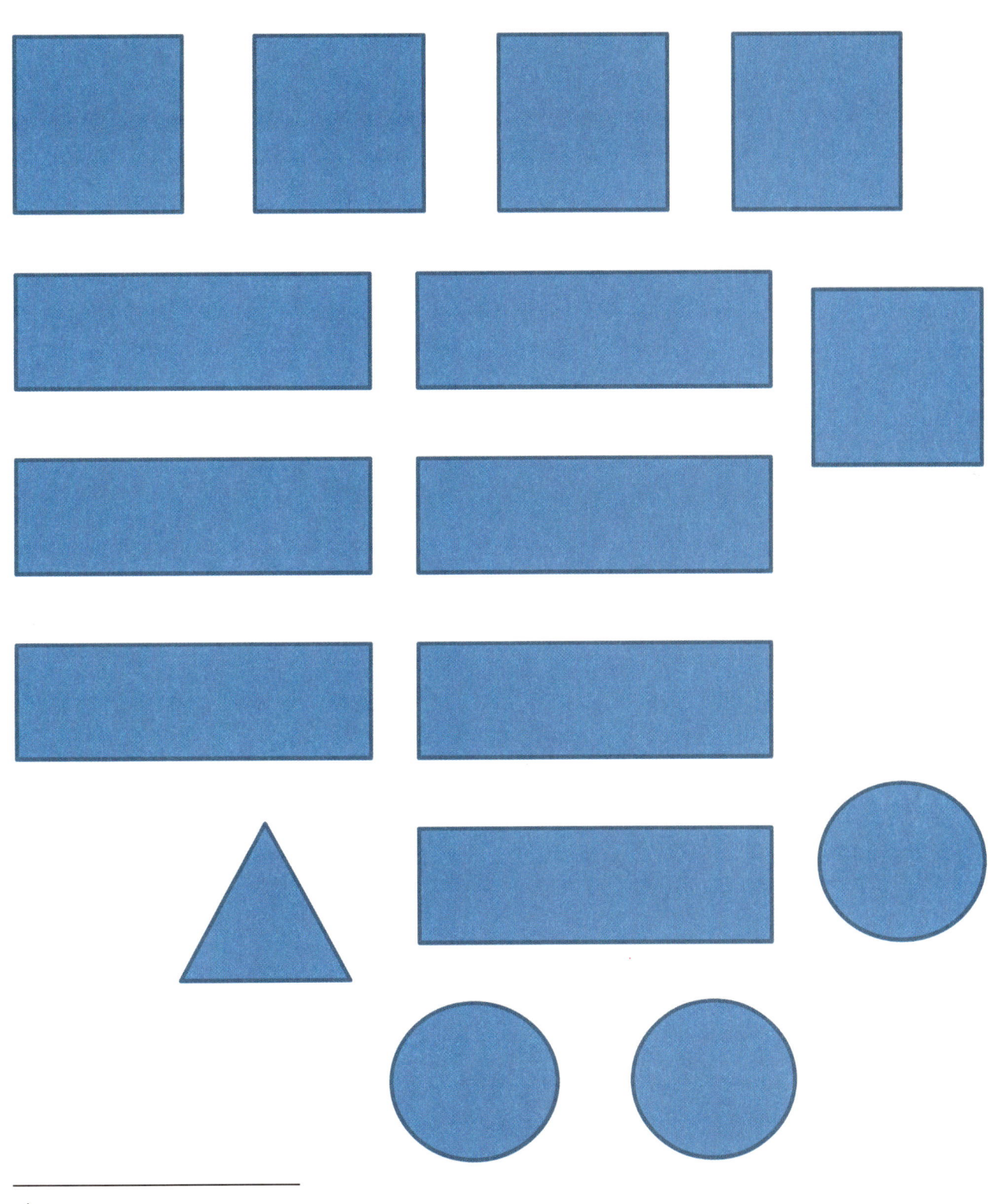

Lesson 21 Template 1   K•3

shapes

Lesson 21: Compare sets informally using *more*, *less*, and *fewer*.

225

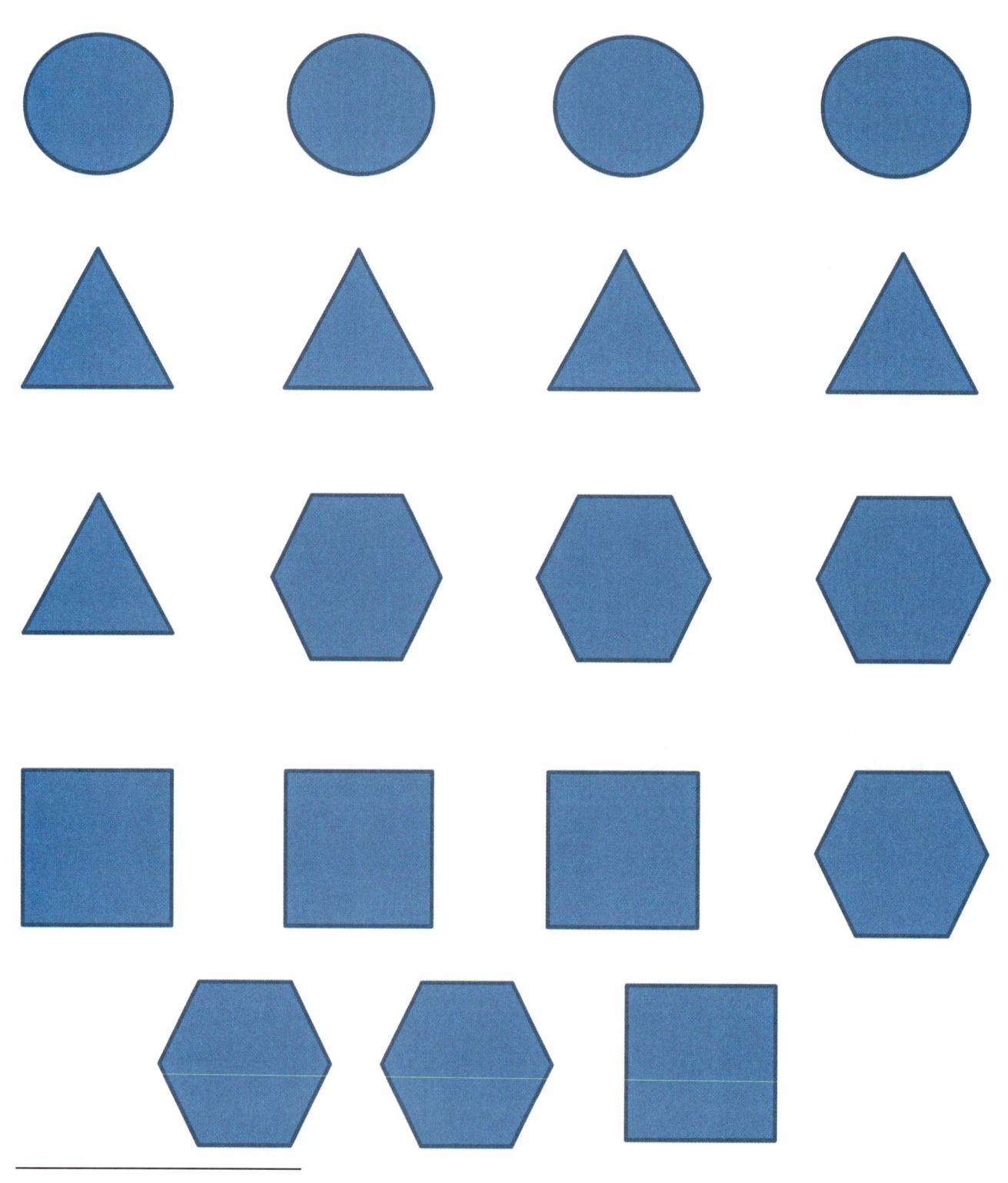

shapes

A STORY OF UNITS   Lesson 21 Template 2 K•3

Name _____ Date _____

Draw a shape to make the sentence true.

There are more _____ than ⬢ .

There are fewer ▲ than _____.

There are fewer _____ than ▬ .

_____
more than, fewer than recording sheet

Lesson 21:   Compare sets informally using *more*, *less*, and *fewer*.

227

Lesson 22

Objective: Identify and create a set that has the same number of objects.

Suggested Lesson Structure

- ■ Fluency Practice   (12 minutes)
- ■ Application Problem   (5 minutes)
- ■ Concept Development   (25 minutes)
- ■ Student Debrief   (8 minutes)

   Total Time   **(50 minutes)**

Fluency Practice (12 minutes)

- Make It Equal **K.CC.6**   (3 minutes)
- Roll and Draw 5-Groups **K.OA.3**   (5 minutes)
- 5-Group Fill-Up **K.OA.4**   (4 minutes)

Make It Equal (3 minutes)

Note: Students visually experience comparison, which is a skill foundational to the work of this module.

Conduct the activity as outlined in Lesson 15.

Roll and Draw 5-Groups (5 minutes)

Note: Observe to see which students erase completely and begin each time from one, rather than drawing more or erasing some to adjust to the new number. By drawing 5-groups, students see numbers as having length in relationship to the five.

Conduct the activity as outlined in Lesson 7. Consider alternating between drawing the 5-groups vertically or horizontally.

5-Group Fill-Up (4 minutes)

Materials: (S) Dice with 6-dot side covered, personal white board

Note: This activity provides students with a head start in terms of learning their partners to ten, thus anticipating the work of the next module.

1. Partner A rolls the dice and draws a corresponding 5-group with Os.
2. Partner B completes the 10 by drawing Xs.
3. Both partners engage in math talk: "I have 3. You drew 7 more to make 10."

A STORY OF UNITS   Lesson 22   K•3

Application Problem (5 minutes)

Materials: (S) 7 linking cubes, small piece of clay

Pretend your linking cubes are little baskets. Use your clay to make as many balls as there are baskets. Check your work by putting a ball in each basket. Do you have just enough? Score 1 point for every basket you made!

Note: The concrete activity of creating an equal set serves as an anticipatory set for today's lesson objective. As you circulate, encourage the students to use the language, "I have the same number of balls as baskets."

Concept Development (25 minutes)

Materials: (S) 10-sided die (or spinner), bag of 20 linking cubes, and bag of 20 pennies

MP.6

T: We are going to play Match My Set today! Let me show you how it works. Student A, please roll the die. What number do you see?
S: 8.
T: I will draw a set of 8 shapes. What shape should I draw, Student A?
S: Circles!
T: (Draw 8 circles on the board.) Now, I will draw as many squares as circles. Then, I'll have **the same number** of squares as circles. (Demonstrate.) How should I check my work?
S: You could count them!
T: Good idea. Count the circles with me.
S: 1, 2, 3, 4, 5, 6, 7, 8.
T: I will write the number 8 under this set. Now, let's count the squares.
S: 1, 2, 3, 4, 5, 6, 7, 8.
T: I will write the number 8 under this set. Do I have the same number of shapes in each set?
S: Yes! They both have 8 shapes!
T: Now, you will play this game with your partner. One of you will roll the die and make the first set with the cubes. Then, the other will make a set of pennies that has the same number of pennies as cubes. When you have made your sets, count each of them to make sure they are the same! The next time, you can switch. (Allow students to play several iterations of the game. Circulate to ensure accuracy in terms of counting and matching.)

> **NOTES ON MULTIPLE MEANS OF REPRESENTATION:**
>
> Scaffold essential terms for English language learners so that they can follow the Application Problem directions and participate fully in the day's lesson. To highlight the concept *just enough*, modeling can be done with two sets of 4 linking cubes. Start with a 4-stick. Match loose linking cubes to the 4-stick one at a time to show *not enough* until all 4 linking cubes have a partner cube to show *just enough*.

> **NOTES ON MULTIPLE MEANS OF ENGAGEMENT:**
>
> Model the steps of the game for students who are below grade level. Play the game with a student while explaining the procedure one step at a time.
>
> "You rolled a 4. Watch me make a set of 4 cubes."
>
> "Now, it's your turn. Use the pennies to make the same number as my cubes."
>
> Count out the sets, if necessary, until the student is able to work with a partner.

EUREKA MATH   Lesson 22:   Identify and create a set that has the same number of objects.

A STORY OF UNITS

Lesson 22 K•3

Problem Set (10 minutes)

Students should do their personal best to complete the Problem Set within the allotted time.

Student Debrief (8 minutes)

Lesson Objective: Identify and create a set that has the same number of objects.

The Student Debrief is intended to invite reflection and active processing of the total lesson experience.

Invite students to review their solutions for the Problem Set. They should check work by comparing answers with a partner before going over answers as a class. Look for misconceptions or misunderstandings that can be addressed in the Debrief. Guide students in a conversation to debrief the Problem Set and process the lesson.

Any combination of the questions below may be used to lead the discussion.

- When you were making the sets with your cubes and pennies, how did you check to make sure that the sets had **the same number** of items?
- What would it mean if you counted 8 in one set and 6 in another?
- What do we have to remember when we are making sets that have the same number of items?
- On the second page of the Problem Set, did your partner draw the correct number of objects to match your set?
- Use the words *the same number* to tell me something about your hands. Could you make a similar sentence about *the same number* for any other part of your body?
- What new (or significant) math vocabulary did we use today to communicate precisely?

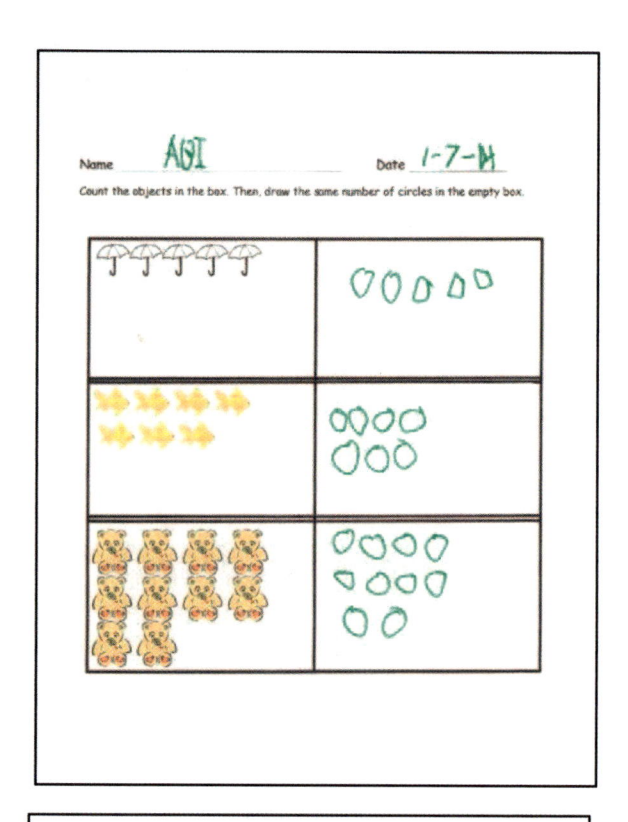

Lesson 22: Identify and create a set that has the same number of objects.

A STORY OF UNITS　　　　　　　　　　　　　　　　Lesson 22 Problem Set　K•3

Name _____ Date _____

Count the objects in the box. Then, draw the same number of circles in the empty box.

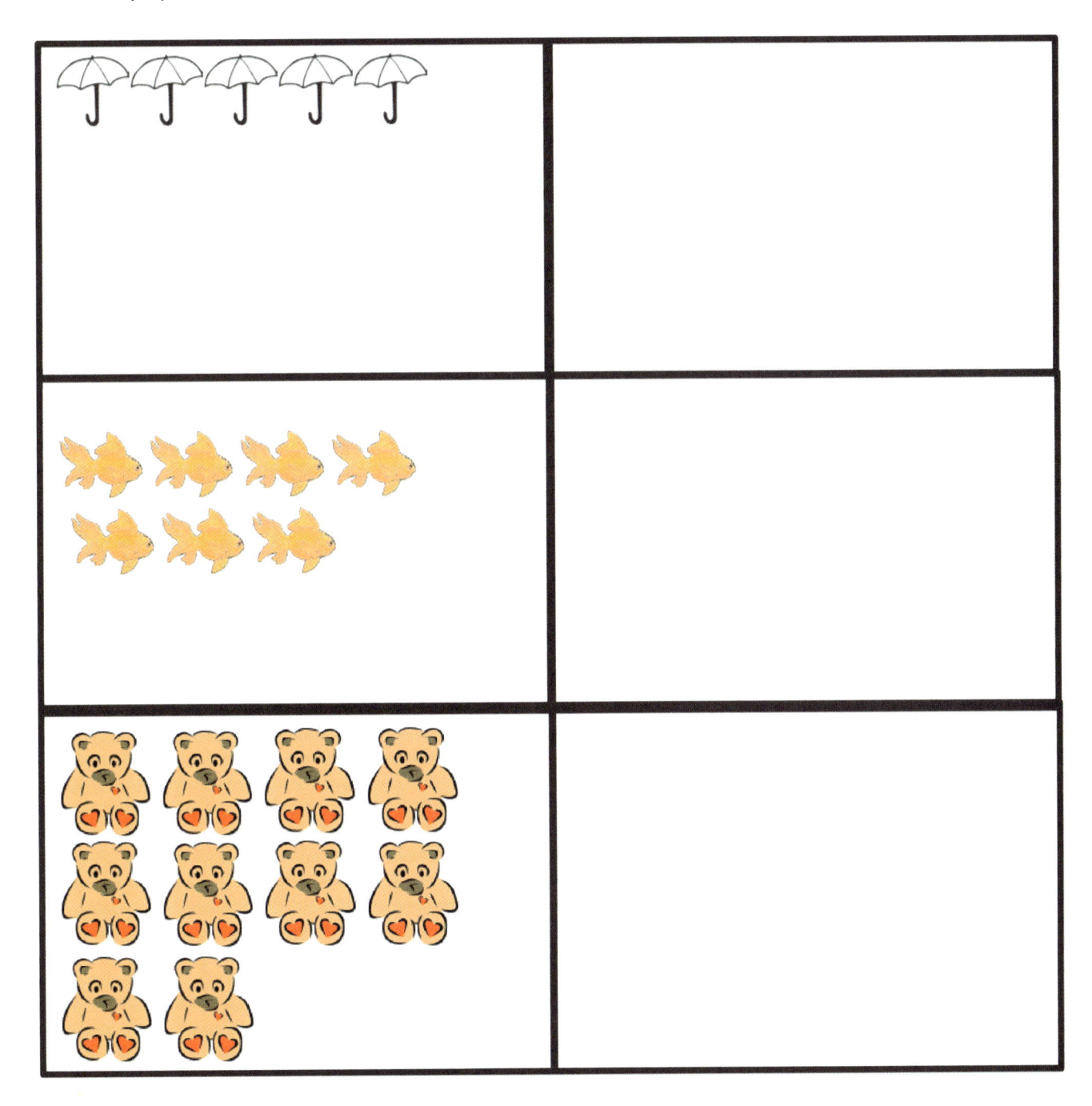

Lesson 22: Identify and create a set that has the same number of objects.

A STORY OF UNITS — Lesson 22 Problem Set — K•3

Draw a set of objects in the first box. Switch papers with a partner. Have your partner draw the same number of objects in the next box.

Lesson 22: Identify and create a set that has the same number of objects.

A STORY OF UNITS Lesson 22 Homework K•3

Name _____ Date _____

Count the birds. In the next box, draw the same number of nests as birds.

Count the houses. In the next box, draw the same number of trees as houses.

Count the monkeys. In the next box, draw the same number of bananas as monkeys.

On the back of your paper, draw some pencils. Then, draw a crayon for each pencil.

Lesson 22: Identify and create a set that has the same number of objects.

Lesson 23

Objective: Reason to identify and make a set that has 1 more.

Suggested Lesson Structure

- **Fluency Practice**   (11 minutes)
- **Application Problem**   (5 minutes)
- **Concept Development**   (26 minutes)
- **Student Debrief**   (8 minutes)
- **Total Time**   **(50 minutes)**

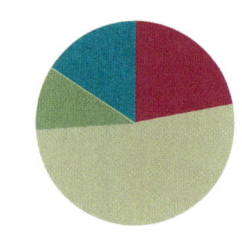

Fluency Practice (11 minutes)

- Show Me 1 More **K.CC.4c**   (4 minutes)
- Roll and Say 1 More **K.CC.4c**   (3 minutes)
- Finish My Sentence (1 More) **K.CC.4c**   (4 minutes)

Show Me 1 More (4 minutes)

Note: Students continue to develop Fluency Practice in terms of describing the pattern of 1 more, preparing them for the current lesson.

Conduct the activity as described in Lesson 19, but focus exclusively on practicing 1 more. Maintain consistency in the language.

Roll and Say 1 More (3 minutes)

Note: This is a reiteration of the previous activity. A different representation (dice in this case), develops flexibility and ensures that students do not become too dependent on finger counting.

Conduct the activity as described in Lesson 13, but focus exclusively on practicing 1 more. Maintain consistency in the language.

Finish My Sentence (1 More) (4 minutes)

Note: The previous fluency activities in this lesson build up to this more abstract version in preparation for today's lesson.

- T: Raise your hand, and wait for the signal for when you can finish this sentence. 3. 1 more is…? (Wait for all hands to go up, and then signal.)
- S: 4.
- T: 4. 1 more is…? (Wait for all hands to go up, and then signal.)
- S: 5.

Variation: After some whole group practice, have students complete this activity with a partner.

Application Problem (5 minutes)

Draw 9 birds. Draw enough worms so that each bird gets one, but also draw 1 extra worm for a snack for later. Use your ruler to match each bird to its worm. How many birds are there? Write the number. How many worms are there? Write the number. Show your picture to a friend.

Note: Creating a set of *enough* but with an *extra one* provides the anticipatory set for today's lesson objective.

Concept Development (26 minutes)

Materials: (S) 10-sided die, bag of 20 linking cubes, bag of 20 pennies per pair

- T: We are going to play another set game today. Let me show you how we will play. Student A, please roll the die. What number do you see?
- S: 4.
- T: I will draw a set of 4. What shape should I draw, Student A?
- S: Triangles!
- T: (Draw 4 triangles on the board.) Now, I need to draw a set of squares that has **1 more** than my set of triangles. How many should I draw? Do you remember how we learned to count *1 more than* with our linking cube stairs a long time ago? We will do that again. Count the triangles with me.
- S: 1, 2, 3, 4.
- T: 4. I will write 4 under this set. What is 1 more?
- S: 1 more is 5.
- T: 4. 1 more is 5. (Draw 5 squares.) I will write the number 5 under this set. Do the sets have the same number?
- S: No! 5 is 1 more than 4.

Model the exercise one more time, having a different student roll the die. Encourage the use of language such as, "6. 1 more is 7. 7. 1 more is 8."

- T: Now, you will play the game with your partner. One of you will roll the die and make the first set with the cubes, and then the other will make a set of pennies that has 1 more than the set of cubes. After you have made your sets, count each of them again to make sure that the set of pennies has 1 more! The next time, you can switch.

Allow students to play several iterations of the game. Circulate to ensure accuracy in terms of counting and matching.

NOTES ON MULTIPLE MEANS OF REPRESENTATION:

To help English language learners participate fully in the lesson, point to visuals of triangles, squares, and other shapes on the word wall as shape names are spoken. If visuals are not posted, add them as a reference for students.

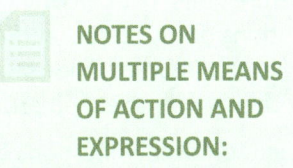

NOTES ON MULTIPLE MEANS OF ACTION AND EXPRESSION:

Scaffold the lesson for students working below grade level by modeling what to do one step at a time. Have one student roll the die. Direct the student's partner to make a set of cubes equal to the number on the die, counting each one. Then, help students make a set of pennies that has 1 more by counting them one at a time. Ask, "Is there 1 more penny than there are cubes?" and so forth until students are able to continue on their own.

Problem Set (10 minutes)

Students should do their personal best to complete the Problem Set within the allotted time.

Note: Before students begin the second page of the Problem Set, encourage the students to think about what a set could look like. Do they look just like their friends'? Do all peanuts, pencils, squirrels, or puppies look identical? Encourage students to draw a set of objects that is diverse. This allows students to find and discuss embedded numbers.

Student Debrief (8 minutes)

Lesson Objective: Reason to identify and make a set that has 1 more.

The Student Debrief is intended to invite reflection and active processing of the total lesson experience.

Invite students to review their solutions for the Problem Set. They should check work by comparing answers with a partner before going over answers as a class. Look for misconceptions or misunderstandings that can be addressed in the Debrief. Guide students in a conversation to debrief the Problem Set and process the lesson.

Any combination of the questions below may be used to lead the discussion.

- In our activity, how did you know how many cubes you needed to use in your set each time?
- How did you know how many pennies should be in the set each time?
- Think about the birds and the worms you drew at the beginning of math today. What could you say about the sets of birds and worms?
- On the Problem Set, what did you do to make sure you drew a set with **1 more**? Talk to your partner about the second page of the Problem Set. Pick one box and talk about the number you rolled and how many objects you drew. (Encourage students to talk about hidden partners, if applicable. For example, how many puppies are playing? How many are eating?)
- What math vocabulary did we use today to communicate precisely?

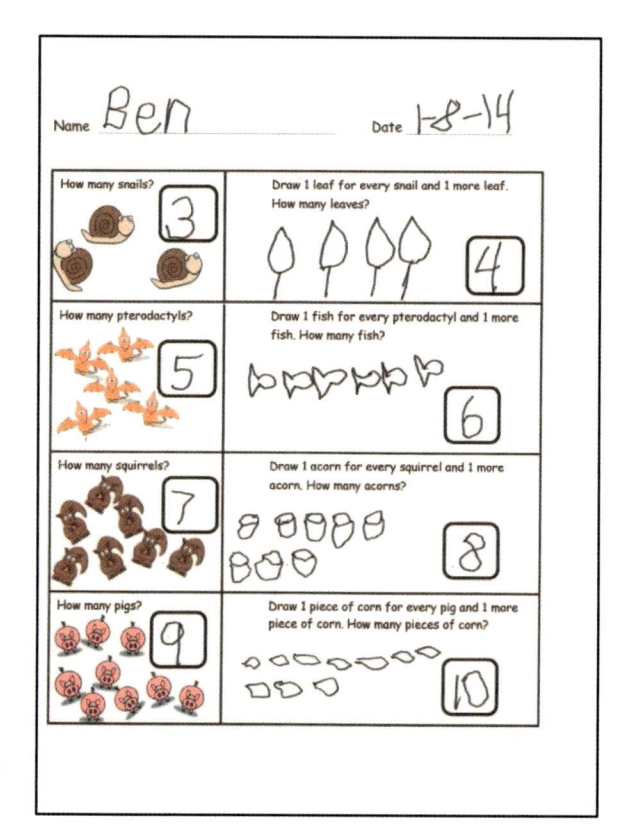

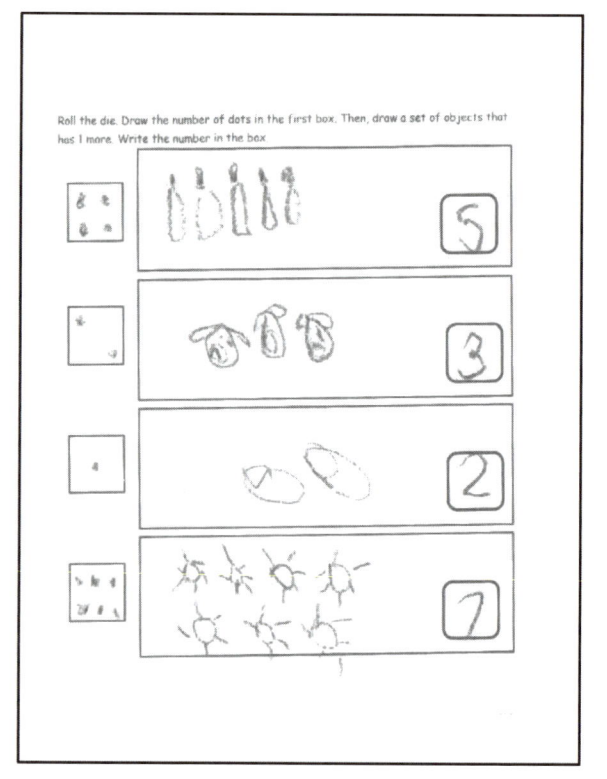

A STORY OF UNITS Lesson 23 Problem Set K•3

Name _____   Date _____

How many snails? ☐	Draw 1 leaf for every snail and 1 more leaf. How many leaves? ☐
How many pterodactyls? ☐	Draw 1 fish for every pterodactyl and 1 more fish. How many fish? ☐
How many squirrels? ☐	Draw 1 acorn for every squirrel and 1 more acorn. How many acorns? ☐
How many pigs? ☐	Draw 1 piece of corn for every pig and 1 more piece of corn. How many pieces of corn? ☐

Lesson 23: Reason to identify and make a set that has 1 more.

A STORY OF UNITS

Lesson 23 Problem Set K•3

Roll the die. Draw the number of dots in the first box. Then, draw a set of objects that has 1 more. Write the number in the box.

238   Lesson 23:   Reason to identify and make a set that has 1 more.

A STORY OF UNITS   Lesson 23 Homework   K•3

Name _____   Date _____

How many cats? ☐

Draw a ball for every cat and 1 more ball.

How many balls? ☐

How many elephants? ☐

Draw a peanut for every elephant and 1 more peanut.

How many peanuts? ☐

Lesson 23:   Reason to identify and make a set that has 1 more.   239

Lesson 24

Objective: Reason to identify and make a set that has 1 less.

Suggested Lesson Structure

- **Fluency Practice** (11 minutes)
- **Application Problem** (5 minutes)
- **Concept Development** (26 minutes)
- **Student Debrief** (8 minutes)
- **Total Time** **(50 minutes)**

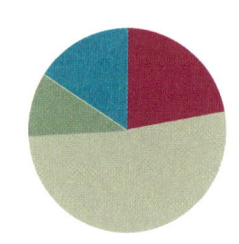

Fluency Practice (11 minutes)

- Show Me 1 Less **K.CC.4c**   (4 minutes)
- Roll and Say 1 Less **K.CC.4c**   (3 minutes)
- Finish My Sentence (1 Less) **K.CC.4c**   (4 minutes)

Show Me 1 Less (4 minutes)

Note: Students continue to develop Fluency Practice in terms of describing the pattern of 1 less, preparing them for the current lesson. This activity echoes the previous lesson's work with 1 more, reinforcing the opposite nature of the concepts.

Conduct the activity as described in Lesson 19, but instead, focus exclusively on practicing 1 less. Maintain consistency in the language.

Roll and Say 1 Less (3 minutes)

Note: This is a reiteration of the previous activity. A different representation (dice in this case), develops flexibility and ensures that students do not become too reliant on finger counting.

Conduct the activity as described in Lesson 13, but focus exclusively on practicing 1 less. Maintain consistency in the language.

Finish My Sentence (1 Less) (4 minutes)

Note: The previous fluency activities in this lesson build up to this more abstract version in preparation for today's lesson.

- T: Raise your hand, and wait for the signal when you can finish this sentence. 5. 1 less is…? (Wait for all hands to go up, and then signal.)
- S: 4.
- T: 4. 1 less is…? (Wait for all hands to go up, and then signal.)
- S: 3.

Variation: After some whole group practice, have students do this activity with a partner.

A STORY OF UNITS   Lesson 24 K•3

Application Problem (5 minutes)

The birds are back! Draw 9 birds. Each of them wants a worm for lunch today except for one—she has become a vegetarian. Draw just enough worms so that each bird who wants one can have one. How many birds did you draw? Write the number. How many worms did you draw? Write the number.

Note: Today's lesson closely mirrors the previous lesson, but the focus is on *1 less* rather than *1 more*. Having the students draw *just enough* worms *except for one* will provide the anticipatory set for the lesson.

> **NOTES ON MULTIPLE MEANS OF ENGAGEMENT:**
>
> Scaffold the Application Problem for students working below grade level by asking questions such as, "How many birds are not vegetarian?" Watch as students draw their birds, count them, and write the number. Ask, "Do you have to draw 9 worms?" Continue questioning until students are successful.

Concept Development (26 minutes)

Materials: (S) 10-sided die, bag of 20 linking cubes, bag of 20 pennies per pair

T: We have one last set game to play! Student A, please roll the die. What did you get?
S: 6.
T: I will draw a set of 6. What shape should I draw, Student A?
S: Hexagons!
T: (Draw 6 hexagons on the board.) Now, I need to draw a set of squares that has **1 fewer than** my set of hexagons. Do you remember how we learned to count *1 less* with our linking cube stairs? We will do that again. Count the hexagons with me.
S: 1, 2, 3, 4, 5, 6.
T: 6. I will write 6 under this set. What is 1 less, or 1 fewer, than 6?
S: 1 less than 6 is 5.
T: 6. 1 less is 5. (Draw 5 squares.) I will write the number 5 under this set. Are the sets the same?
S: No! 6 is 1 more than 5.

> **NOTES ON MULTIPLE MEANS OF ENGAGEMENT:**
>
> English language learners are often shy about producing language. Practice saying, "6. 1 less than 6 is 5," with the whole class. Vary the choral response so that the boys try it alone, then the girls, then the left side of the room, then the right, etc. Practicing helps English language learners gain confidence in producing language.

MP.2

Model the exercise one more time, having a different student roll the die. Encourage the use of language such as, "9. 1 less is 8. 8. 1 less is 7."

T: Now, you will play the game with your partner. One of you will roll the die and make the first set with the cubes, and then the other will make a set of pennies that has 1 fewer than the set of cubes. After you make your sets, count each of them again to make sure that the number of pennies is one less! Next time, you can switch. (Allow students to play several iterations of the game. Circulate to ensure accuracy in terms of counting and matching.)

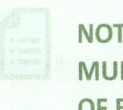

Lesson 24: Reason to identify and make a set that has 1 less.

Lesson 24

Problem Set (10 minutes)

Students should do their personal best to complete the Problem Set within the allotted time.

Student Debrief (8 minutes)

Lesson Objective: Reason to identify and make a set that has 1 less.

The Student Debrief is intended to invite reflection and active processing of the total lesson experience.

Invite students to review their solutions for the Problem Set. They should check work by comparing answers with a partner before going over answers as a class. Look for misconceptions or misunderstandings that can be addressed in the Debrief. Guide students in a conversation to debrief the Problem Set and process the lesson.

Any combination of the questions below may be used to lead the discussion.

- When you were playing the game, how did you know how many pennies needed to be in a set?
- If your partner made a set of 5 pennies, how many cubes would you have put in a set?
- What if he had made a set of 9 pennies? How many cubes would you have put in the set?
- On the Problem Set, how did you know how many chicks there were? How did you know how many worms to draw?
- What math vocabulary did we use today to communicate precisely?
- Think about the birds and the worms you drew at the beginning of math today. What could you say about the sets of birds and worms?

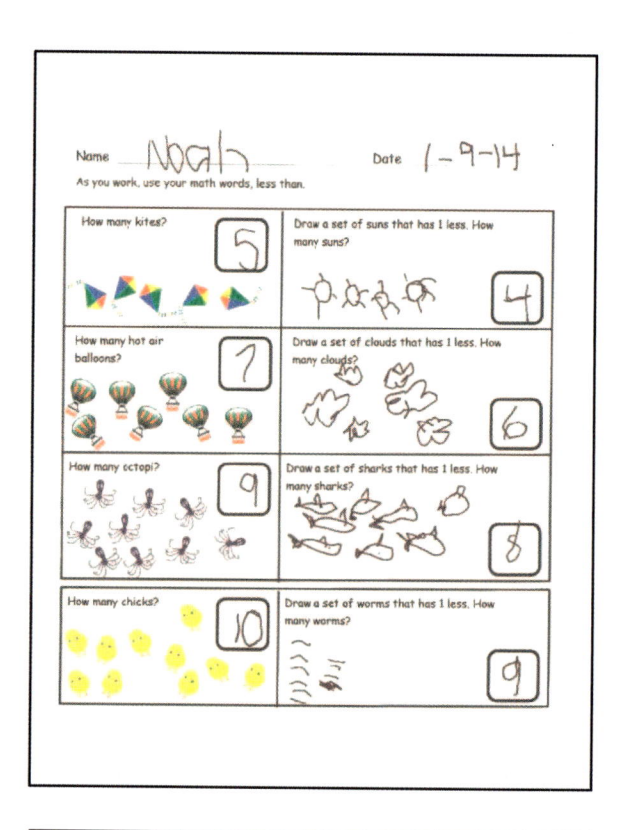

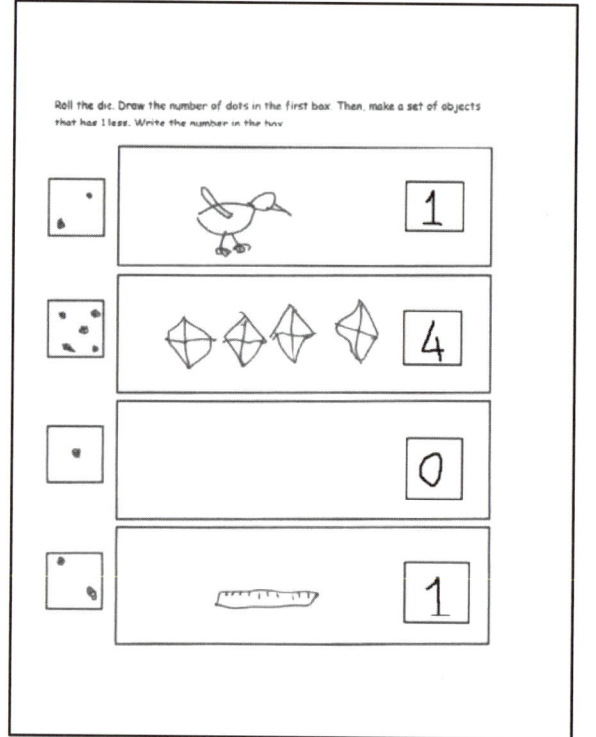

A STORY OF UNITS

Lesson 24 Problem Set K•3

Name _____ Date _____

As you work, use your math words *less than*.

How many kites? ☐	Draw a set of suns that has 1 less. How many suns? ☐
How many hot air balloons? ☐	Draw a set of clouds that has 1 less. How many clouds? ☐
How many octopi? ☐	Draw a set of sharks that has 1 less. How many sharks? ☐
How many chicks? ☐	Draw a set of worms that has 1 less. How many worms? ☐

Lesson 24: Reason to identify and make a set that has 1 less.

A STORY OF UNITS

Lesson 24 Problem Set K•3

Roll the die. Draw the number of dots in the first box. Then, make a set of objects that has 1 less. Write the number in the box.

244 Lesson 24: Reason to identify and make a set that has 1 less.

Lesson 24 Homework K•3

Name _____ Date _____

Count the set of objects, and write how many in the box.

Draw a set of circles that has 1 less, and write how many in the box. As you work, use your math words *less than*.

A STORY OF UNITS

Mathematics Curriculum

GRADE K • MODULE 3

Topic G
Comparison of Numerals

K.CC.6, K.CC.7, K.CC.4c

Focus Standards:	K.CC.6	Identify whether the number of objects in one group is greater than, less than, or equal to the number of objects in another group, e.g., by using matching and counting strategies. (Include groups with up to ten objects.)
	K.CC.7	Compare two numbers between 1 and 10 presented as written numerals.
Instructional Days:	4	
Coherence -Links from:	GPK–M4	Comparison of Length, Weight, Capacity, and Numbers to 5
-Links to:	G1–M3	Ordering and Comparing Length Measurements as Numbers

Topic G is a bridge that enables students to compare numerals by connecting number to length. In Lessons 25 and 26, students work with linear configurations to match and count to see that "7 is more than 3, 3 is less than 7, and 5 is equal to 5."

In Lessons 26 and 27, students look for and find strategies to compare sets of objects in various configurations.

Finally, in Lesson 28, students visualize quantities as they compare numerals without using materials, a skill that will be fine-tuned throughout the balance of the Kindergarten year.

A Teaching Sequence Toward Mastery of Comparison of Numerals
Objective 1: Match and count to compare a number of objects. State which quantity is more. (Lesson 25)
Objective 2: Match and count to compare two sets of objects. State which quantity is less. (Lesson 26)
Objective 3: Strategize to compare two sets. (Lesson 27)
Objective 4: Visualize quantities to compare two numerals. (Lesson 28)

Lesson 25

Objective: Match and count to compare a number of objects. State which quantity is more.

Suggested Lesson Structure

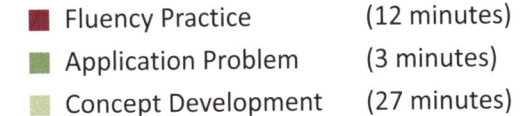

- Fluency Practice   (12 minutes)
- Application Problem   (3 minutes)
- Concept Development   (27 minutes)
- Student Debrief   (8 minutes)

Total Time   **(50 minutes)**

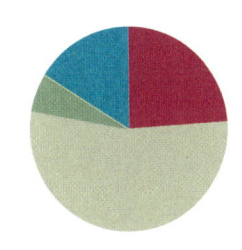

Fluency Practice (12 minutes)

- Beat Your Score! **K.CC.4b**   (12 minutes)

Beat Your Score! (12 minutes)

Materials: (S) 2 copies of Count and Circle How Many (Lesson 20 Sprint)

Note: The purpose of this activity is to help students become accustomed to the full Sprint routine while completing a task involving relatively simple concepts (hence the reuse of a Sprint from Lesson 20). This activity builds confidence and enthusiasm for Sprints.

- T: It's time for a Sprint! (Briefly recall previous Sprint preparation activities, and distribute Sprints face down.) Take out your pencil and one crayon, any color.
- T: On your mark, get set, go!
- S: (Work.)
- T: (Ring the bell, or give another signal for students to stop. Although it will not be necessary to time the students in this short practice Sprint, be sure to give the stop signal before students finish so they do not develop the expectation of finishing every time.) Pencils up!
- T: Pencils down, crayons up!
- T: It's time to check answers. What do you do if the answer is right?
- S: Circle it. (Circling correct answers instead of crossing out wrong ones avoids stigmatization.)
- T: What do you say?
- S: Yes!
- T: We'll begin with the hearts. Ready? 1.
- S: Yes!

Proceed through the checking answers procedure, as in Lesson 21.

T: Kindergarteners, do you ever wish you had more time? Another chance to do even better?

S: Yes.

T: Before we try again, let's get our mind and body ready to work hard with an exercise. Stand up and push in your chairs. Let's do jumping jacks while counting to 10. Ready?

S: 1, 2, 3, …, 10. (Count while doing jumping jacks.)

T: Hands on your hips. Twist slowly, counting down from 10. Ready? (While students exercise, distribute the second set of Sprints, which is the same as the first.)

S: 10, 9, 8, …, 1. (Count while twisting.)

T: Have a seat. Pencils up. Do you remember the number you got the first time?

S: Yes.

T: See if you can beat your own score! Race against yourself! On your mark, get set, go!

Students work on the Sprint for a second time. Perhaps give an additional three to five seconds to help students beat their first score. Give the signal to stop, reiterating that it is okay not to finish. Continue to emphasize that the goal is simply to do better than the first time. Proceed through the checking answers procedure with more enthusiasm than ever. Then, facilitate a comparison of Sprint A to Sprint B. Because students are still developing understanding of the concept of more, it may be necessary to circulate and facilitate the comparison, either visually or numerically.

T: Stand up if you beat your score.

T: You worked so hard, and I am so proud of you! Let's celebrate (e.g., congratulate each other, give three pats on the back, shake hands, have a parade).

Variation: Allow students to finish, but provide an early finisher activity to do on the back.

Application Problem (3 minutes)

Materials: (S) Bag of 10 pennies, bag of 8 linking cubes

Note: This Application Problem introduces the comparing of sets of objects in linear configurations, serving as an anticipatory set for the lesson.

Put your pennies in a row. Now, put one linking cube on top of each penny. Are there enough cubes to cover each penny? Talk to your friend about which has more, the set of cubes or the set of pennies.

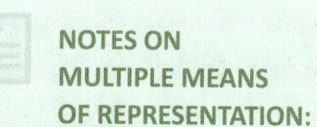

> **NOTES ON MULTIPLE MEANS OF REPRESENTATION:**
>
> Model the Application Problem for English language learners. Show what to do by placing a linking cube on top of a penny while speaking the instruction. Model how to tell a partner which set has more: "I have more pennies than linking cubes because two pennies are not covered."

Concept Development (27 minutes)

Materials: (T) White board and markers, shapes (Lesson 21 Template 1), cut out and placed in scatter arrangements on the board

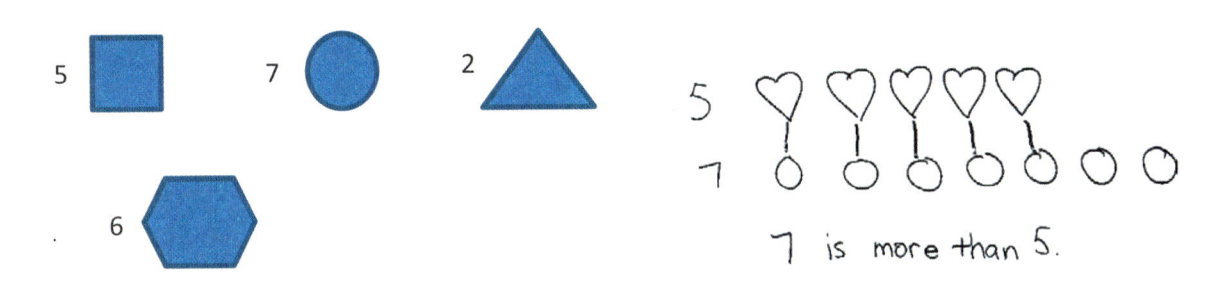

T: What do you notice on the board today?
S: We have lots of shapes.
T: Do you remember the names of the shapes?
S: There are triangles and hexagons. We have circles and squares.
T: We've been talking lately about sets that have *more than* and *less than*. Today, we are going to talk about ways to organize our groups of shapes so that it is easier to tell which has more.
T: Which has more, the circles or triangles?
S: There are more circles than triangles.
T: How did you know so fast?
S: I could just see there were lots more. → Yeah, I didn't have to count because there are circles all over the place and just 2 triangles. → I didn't count the circles, but I could see there were more than 2.
T: That makes sense, but what about the squares and the hexagons? Right now, it is hard for me to guess which has more. It isn't so easy to just see. Do you have any ideas?
S: (Discuss.)
T: (Guide the discussion so that students remember how they worked with the coins and cubes in previous lessons.)
S: Let's line them up!
T: I can move our shapes. I will put the squares in a row and the hexagons in a row just underneath. (Demonstrate.) Now, what do you notice?
S: The hexagon line is longer. → The hexagons are bigger. Maybe there are more, but I can't tell.

MP.2

NOTES ON MULTIPLE MEANS OF REPRESENTATION:

Scaffold the lesson for students working below grade level and those having trouble grasping the concept of one-to-one correspondence by matching hexagons and squares one at a time. "One hexagon. Let's count one square. Two hexagons. Two squares." etc. Once students get the idea, move on to counting one set with more members than the other.

Lesson 25: Match and count to compare a number of objects. State which quantity is more.

A STORY OF UNITS   Lesson 25   K•3

MP.2

T: We can show which set has more. Let's draw a line between the first hexagon and the first square. (Demonstrate.) Now, let's match the second hexagon with the second square. (Continue until all hexagons are matched.) Each of our hexagons has a partner in the other set. What do you notice now?

S: There's a square left over.

T: I wonder if we could count them to find out which has more. Let's count the hexagons and write that number at the end. 1, 2, 3, 4, 5, 6. Now, let's count the squares. 1, 2, 3, 4, 5.

T: Let's write that number, too. (Write the number.) What do you notice?

T: Look at the numbers at the ends of the lines. There are 6 hexagons and 5 squares. 6 is more than 5. Repeat with me.

S: 6 is more than 5.

T: Here is a question to ask your partner, "Partner, which is more, 6 or 5?" What will your partner say?

S: 6 is more than 5.

T: Take turns, and ask your partner the question.

Repeat activity several times, using various combinations of shapes. Model the linear configuration and one-to-one correspondence each time. Have the students work with their own drawings, representing the shapes as soon as they are ready. They should be able to line things up and match them independently.

Problem Set (10 minutes)

Students should do their personal best to complete the Problem Set within the allotted time.

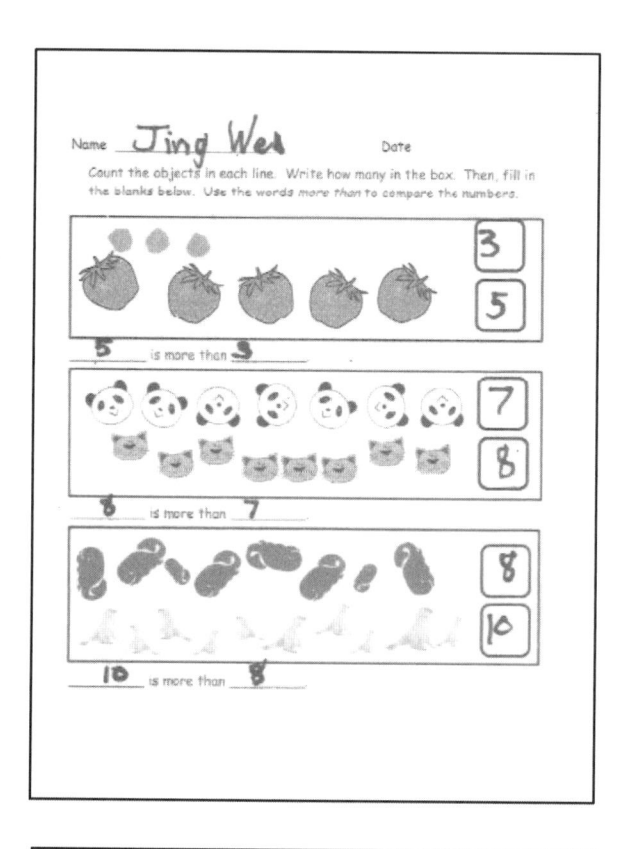

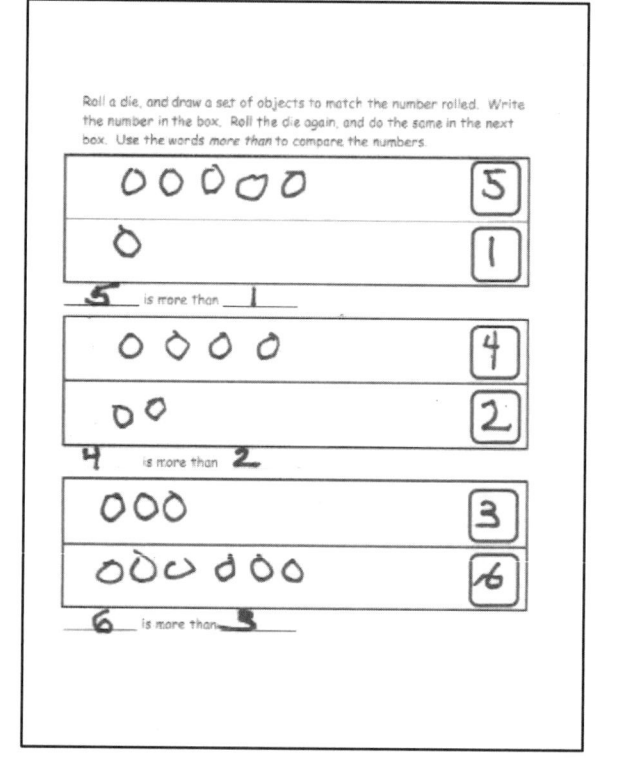

250   Lesson 25:   Match and count to compare a number of objects. State which quantity is more.

Student Debrief (8 minutes)

Lesson Objective: Match and count to compare a number of objects. State which quantity is more.

The Student Debrief is intended to invite reflection and active processing of the total lesson experience.

Invite students to review their solutions for the Problem Set. They should check work by comparing answers with a partner before going over answers as a class. Look for misconceptions or misunderstandings that can be addressed in the Debrief. Guide students in a conversation to debrief the Problem Set and process the lesson.

Any combination of the questions below may be used to lead the discussion.

- How did you organize your shapes to help you know which had more?
- Can you tell by lining up the shapes which has more? How or how not?
- On the Problem Set, how did you know which set had more? Fewer?
- On the second page of the Problem Set, you compared two numbers. Did anyone roll the same number to compare? What did you do?
- What math vocabulary did we use today to communicate precisely? How did the Application Problem connect to today's lesson?

Name _____   Date _____

Count the objects in each line. Write how many in the box. Then, fill in the blanks below. Use the words *more than* to compare the numbers.

_____ is more than _____.

_____ is more than _____.

_____ is more than _____.

A STORY OF UNITS

Lesson 25 Problem Set K•3

Roll a die, and draw a set of objects to match the number rolled. Write the number in the box. Roll the die again, and do the same in the next box. Use the words *more than* to compare the numbers.

_____ is more than _____.

_____ is more than _____.

_____ is more than _____.

Lesson 25: Match and count to compare a number of objects. State which quantity is more.

253

A STORY OF UNITS

Lesson 25 Homework K•3

Name _____ Date _____

Count the objects in each line. Write how many in the box. Then, fill in the blanks below.

_____ is more than _____.

_____ is more than _____.

_____ is more than _____.

254   Lesson 25:   Match and count to compare a number of objects. State which quantity is more.

Lesson 26

Objective: Match and count to compare two sets of objects. State which quantity is less.

Suggested Lesson Structure

- ■ Fluency Practice   (12 minutes)
- ■ Application Problem   (5 minutes)
- ■ Concept Development   (25 minutes)
- ■ Student Debrief   (8 minutes)
- **Total Time**   **(50 minutes)**

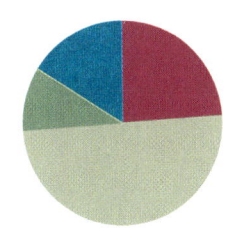

Fluency Practice (12 minutes)

- Matching Fingertips One-to-One **K.CC.6**   (4 minutes)
- Dot Cards of 6 **K.CC.2**   (4 minutes)
- Say Ten Push-Ups **K.NBT.1**   (4 minutes)

Matching Fingertips One-to-One (4 minutes)

Note: This exercise allows students to practice one-to-one matching at the concrete level, preparing them to draw lines to match one-to-one pictorially in this lesson.

Conduct the activity as outlined in Lesson 17.

Dot Cards of 6 (4 minutes)

Materials: (T/S) Dot cards of 6 (Lesson 13 Fluency Template)

Note: Reviewing 6, 7, 8, and 9 is essential in anticipating the work of the next module. While compositions of 5 have been well established at this point, numbers 6 through 9 prove challenging.

Conduct the activity as outlined in Lesson 13.

Say Ten Push-Ups (4 minutes)

Note: This activity extends students' understanding of numbers to 10 in anticipation of working with teen numbers. Conduct the activity as outlined in Lesson 1. Continue to 20 (2 ten, or 10 and 10).

A STORY OF UNITS　　　　　　　　　　　　　　　　　　　　　　　　　　　　　Lesson 26　K•3

Application Problem (5 minutes)

Draw how many people are sitting at your table. Draw them in a row or line. Now, draw to show how many pencils are at your table. Draw them in a row or line. Draw lines to match each person to one pencil. Remember, each one gets only one partner! Are there more pencils or people? Show your work to your partner.

Note: Use this exercise to observe which students demonstrate the concept of one-to-one correspondence. This problem also serves as a review before the *less than* extension of today's objective.

> **NOTES ON MULTIPLE MEANS OF ENGAGEMENT:**
>
> Support English language learners and students working below grade level by asking questions that scaffold an understanding of the Application Problem. For instance, ask, "How many people are at your table? Draw that many faces." And then, "How many pencils are at your table? Draw that many pencils." Finally, ask while pointing from the face to the pencil, "Can you draw a line to connect one face to one pencil?"

Concept Development (25 minutes)

Materials: (T) White board and markers, shapes (Lesson 21 Template 1) cut out and placed in scatter arrangements on the board

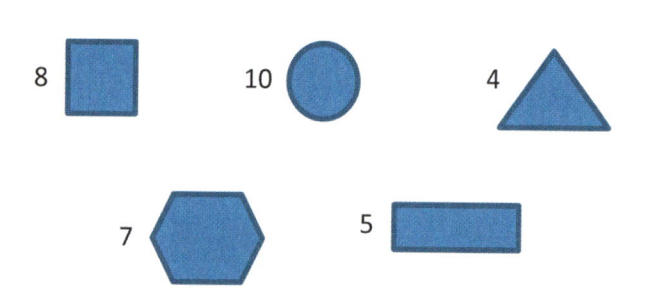

T: In Lesson 25, we talked about how to organize our counting and comparing when we had groups of things. What do you remember?
S: We made lines of things. → We counted them. → We matched them up to find out which had more.
T: We are going to work on more of this today. Look at the shapes on the board. How can we quickly find out if there are more circles or squares?
S: We can line them up.
T: Yes, we can line them up and match them with partners. What if we put them in towers like your linking cubes? What if we put them in columns? Will that still work?
S: Yes!
T: Let's try. (Place circles and squares in columns.) Now, what do I need to remember? How do I match them?
S: Each shape gets only one partner!

Lesson 26: Match and count to compare two sets of objects. State which quantity is less.

T: Does it matter which shape is bigger when I am matching?
S: No.
T: I will draw lines between the partners. (Demonstrate.) What did we discover?
S: There are more circles! → There are leftovers.
T: Count the circles.
S: 1, 2, 3, 4, 5, 6, 7, 8, 9, 10.
T: Now, let's count the squares.
S: 1, 2, 3, 4, 5, 6, 7, 8.
T: Let's write the numbers above each column.
T: Compare the numbers!
T: Just like we did in the other lesson, let's question our partner. Today, let's use the word *less*. Who thinks they have a good question?
S: What is less? → Which is less? → Which number is less, 10 or 8? → Which is less, 8 or 10?
T: Those questions got better and better. Let's use this one: "Which number is less, 10 or 8?" What will your partner say?
S: 8 is less than 10.
T: Great. Begin your interview.

> **NOTES ON MULTIPLE MEANS OF ENGAGEMENT:**
>
> English language learners benefit from hearing and seeing sentence starters like "___ is less than ____," which they can refer to as they do their work. Model the complete sentence, for example, "8 circles is less than 2 triangles," while pointing to the matching visuals.

Repeat with several different combinations of shapes, emphasizing the *less than* language in both the set and number comparisons. Model the one-to-one correspondence carefully. Have the students work with their own drawings when they are ready. They should be able to line things up and match them independently.

T: We will work on this more in our Problem Set.

Problem Set (10 minutes)

Students should do their personal best to complete the Problem Set within the allotted time.

Student Debrief (8 minutes)

Lesson Objective: Match and count to compare two sets of objects. State which quantity is less.

The Student Debrief is intended to invite reflection and active processing of the total lesson experience.

Invite students to review their solutions for the Problem Set. They should check work by comparing answers with a partner before going over answers as a class. Look for misconceptions or misunderstandings that can be addressed in the Debrief. Guide students in a conversation to debrief the Problem Set and process the lesson.

Any combination of the questions below may be used to lead the discussion.

- When we were lining up the shapes on the board to compare the sets, did it matter if we made rows or columns?
- What is the most important thing to remember when lining up shapes? Why does each shape get only one partner?
- What new (or significant) math vocabulary did we use today to communicate precisely?
- How did the Application Problem connect to today's lesson?

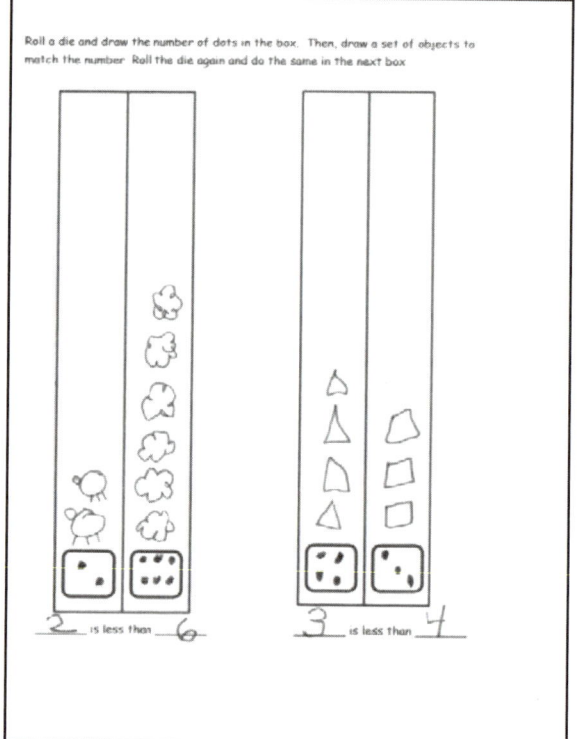

A STORY OF UNITS

Lesson 26 Problem Set   K•3

Name _____   Date _____

Count the objects in each line. Write how many in the box. Then, fill in the blanks below. Say your words *less than* out loud as you work.

_____ is less than _____.

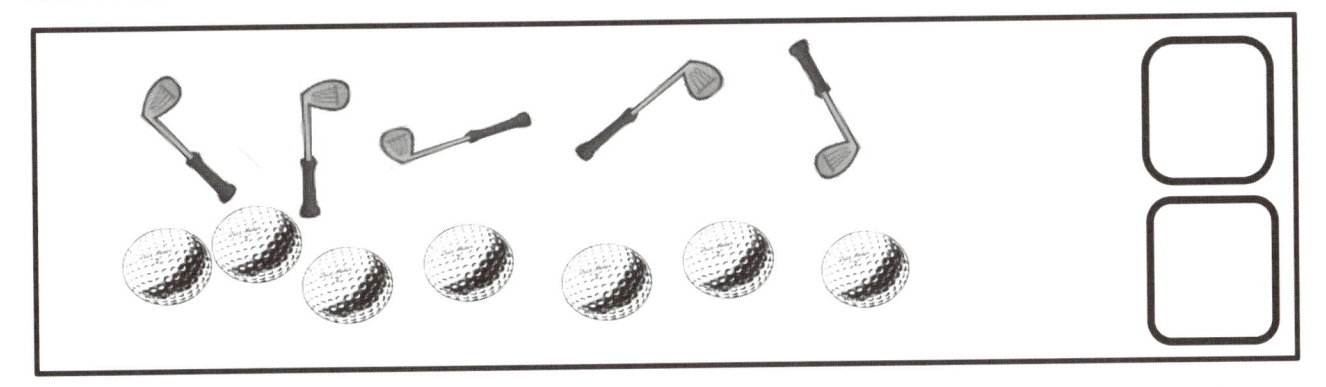

_____ is less than _____.

_____ is less than _____.

Lesson 26: Match and count to compare two sets of objects. State which quantity is less.

Roll a die, and draw the number of dots in the box. Then, draw a set of objects to match the number. Roll the die again, and do the same in the next box.

_____ is less than _____.

_____ is less than _____.

A STORY OF UNITS Lesson 26 Homework K•3

Name _____ Date _____

Count the objects in each line. Write how many in the box. Then, fill in the blanks below.

_____ is less than _____.

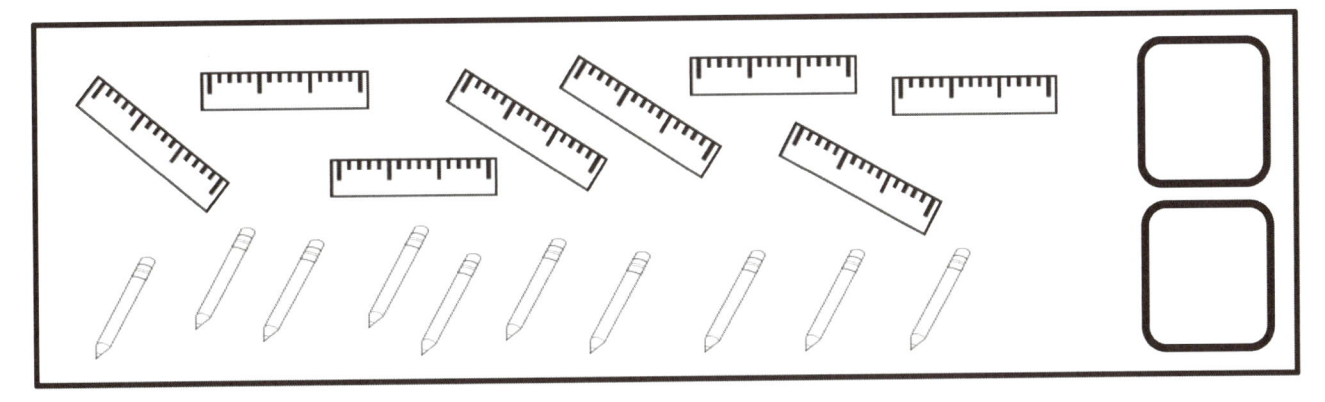

_____ is less than _____.

_____ is less than _____.

Lesson 26: Match and count to compare two sets of objects. State which quantity is less.

Lesson 27

Objective: Strategize to compare two sets.

Suggested Lesson Structure

- **Fluency Practice** (11 minutes)
- **Application Problem** (5 minutes)
- **Concept Development** (26 minutes)
- **Student Debrief** (8 minutes)

Total Time **(50 minutes)**

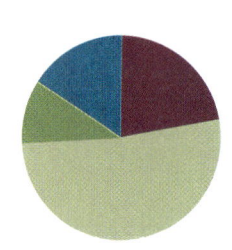

Fluency Practice (11 minutes)

- How Many Are Hiding? **K.OA.4** (4 minutes)
- Hidden Numbers **K.OA.3** (4 minutes)
- Show Me Taller and Shorter **K.MD.1** (3 minutes)

How Many Are Hiding? (4 minutes)

Note: Partners to ten is foundational with respect to development of ten as a unit. Starting early and practicing frequently facilitates automaticity.

- T: How many fingers do you have on two hands?
- S: 10.
- T: Show me 9, piano style, like this. (Demonstrate fingers the Math Way, palms down, flat on the table.)
- T: How many fingers are hiding?
- S: 1.
- T: Let that finger come out now. 9 and 1 make...?
- S: 10.
- T: Now, show me 8.
- T: How many fingers are hiding?
- S: 2.
- T: Let those fingers come out now. 8 and 2 make...?
- S: 10.

Work through all of the combinations of 10.

Hidden Numbers (4 minutes)

Materials: (S) Hidden numbers mat (Lesson 3 Fluency Template)

Note: Finding embedded numbers anticipates the work of Module 4 by developing part–whole thinking.

Conduct the activity as described in Lesson 3, but this time, guide students to find hidden numbers within a group of 6. Look for opportunities to compare sets within the larger group. Encourage students to use the newly acquired vocabulary of *more, less*, and the *same as*. Guide students to say, "6 is 4 and 2, but 4 is more than 2." Or, "6 is 3 and 3. Hey, that's the same number!"

Show Me Taller and Shorter (3 minutes)

Materials: (T) Ruler, pencil

Note: Recalling this vocabulary prepares students for the Concept Development activities in this lesson.

Conduct the activity as described in Lesson 2.

Application Problem (5 minutes)

Materials: (S) Pattern blocks, small bucket per pair

Work with a partner. Take one handful of pattern blocks out of the bucket. Let your partner do the same. Compare your handfuls of pattern blocks. Who has more? How do you know? Put the blocks back, and try the game again.

Note: Circulate during this time to observe student strategies for comparing the sets of blocks. Do the students line them up? Do they match them in pairs? Do they count them? Gather information about their existing strategies to guide your discussions in today's lesson.

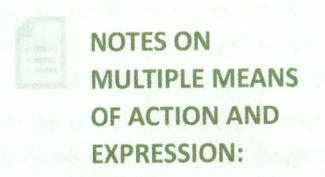

> **NOTES ON MULTIPLE MEANS OF ACTION AND EXPRESSION:**
>
> Extend learning for students working above grade level by challenging them to explain, either orally or in writing, how they knew who had more pattern blocks. Ask them to think of another strategy they can use to determine who has more pattern blocks.

Lesson 27: Strategize to compare two sets.

Concept Development (26 minutes)

Materials: (T) 2 sets of student materials (S) 10-sided die, bag of 10 linking cubes, bag of 10 beans, bag of 10 pennies, bag of 10 counters per pair

T: We are going to do some more comparing activities together, and then you will repeat them with your partner. Watch carefully. Student A, please come up to help.

T: I am going to roll the die and take that many cubes out of the bag. You do the same. (Demonstrate.) What would be a simple way to see who has more?

S: Make towers!

T: (Demonstrate.) Student A, how many cubes are in your tower?

S: 6.

T: I have 8. Let's see whose tower is taller. Which is more, 8 or 6?

S: 8.

T: 8 is more than 6. Use your words.

S: 8 is more than 6.

T: Now, you and your partner try the game. (Allow time for comparison and discussion. Continue to encourage the language of *more than* and *less than*.)

T: Put the cubes away, and watch our next game. Student B, please come up to help. Student B and I will each take some pennies out of our bag. (Demonstrate.) How can we see who has more?

S: Line them up!

T: We will make rows of our pennies. (Demonstrate.) Student A, how many pennies do you have?

S: 9.

T: I have 3. Let's make pairs, and then move our pennies. (Demonstrate.) Who has fewer?

S: You do! You only had 3.

T: 3 is less than 9. Use your words.

S: 3 is less than 9.

T: Thank you, Student B! You and your partner can play the game now. Line up your pennies each time to find out who has more. (Allow time for comparison and discussion.)

T: Put your pennies away. Take out your bag of beans. Roll the die to find out how many beans will be in your set. Compare your set with your partner's. Who has more? How do you know? (Circulate during the lesson to observe strategies of comparison. Encourage students to use multiple strategies and to use *more than* and *less than* vocabulary in their discussions.)

> **NOTES ON MULTIPLE MEANS OF ENGAGEMENT:**
>
> Ask students to verbalize who has more as they take turns every time they play the game. For example, "I have 8 cubes, and you have 3 cubes; 8 is more than 3." Or, "I have 4 pennies, and you have 7 pennies; 4 is less than 7." English language learners benefit from the practice and can be easily observed as to which students might be confused between *more* and *less*.

MP.6

A STORY OF UNITS   Lesson 27   K•3

Problem Set (10 minutes)

Students should do their personal best to complete the Problem Set within the allotted time.

Student Debrief (8 minutes)

Lesson Objective: Strategize to compare two sets.

The Student Debrief is intended to invite reflection and active processing of the total lesson experience.

Invite students to review their solutions for the Problem Set. They should check work by comparing answers with a partner before going over answers as a class. Look for misconceptions or misunderstandings that can be addressed in the Debrief. Guide students in a conversation to debrief the Problem Set and process the lesson.

Any combination of the questions below may be used to lead the discussion.

- How did we compare our sets with the linking cubes? What is another way we could have compared them?
- What do you think was the easiest way to find out which bean set had more? Would you do the same thing to find out which set had fewer beans?
- When do you need to count to see which set has more or less?
- When might we compare numbers in our lives?
- What math vocabulary did we use today to communicate precisely?
- How did the Application Problem connect to today's lesson?

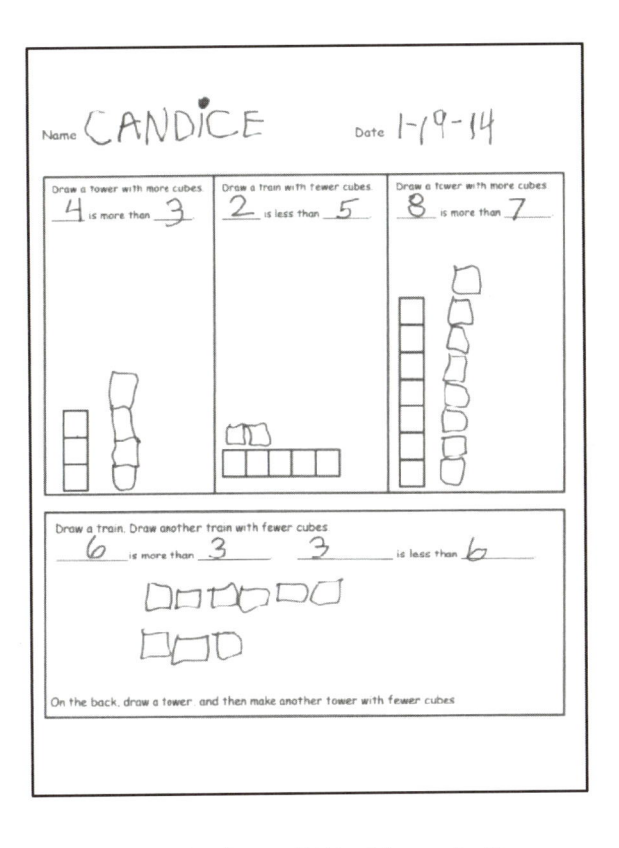

Lesson 27:   Strategize to compare two sets.

A STORY OF UNITS　　　　　　　　　　　　　　　　　　　Lesson 27 Problem Set　K•3

Name _____　　　　Date _____

| Draw a tower with more cubes.

___ is more than ___. | Draw a train with fewer cubes.

___ is less than ___. | Draw a tower with more cubes.

___ is more than ___. |

Draw a train. Draw another train with fewer cubes.

_____ is more than _____.　_____ is less than _____.

266　Lesson 27:　Strategize to compare two sets.

A STORY OF UNITS Lesson 27 Homework K•3

Name _____ Date _____

Draw a tower with more cubes.

_____ is more than _____.

Draw a tower with fewer cubes.

_____ is more than _____.

_____ is less than _____.

Draw a train with more cubes.

_____ is more than _____.

_____ is less than _____.

On the back, draw a tower. Draw another tower that has more cubes.

_____ is more than _____.   _____ is less than _____.

Lesson 27:   Strategize to compare two sets.

267

Lesson 28

Objective: Visualize quantities to compare two numerals.

Suggested Lesson Structure

- Fluency Practice   (12 minutes)
- Application Problem   (5 minutes)
- Concept Development   (25 minutes)
- Student Debrief   (8 minutes)
- **Total Time**   **(50 minutes)**

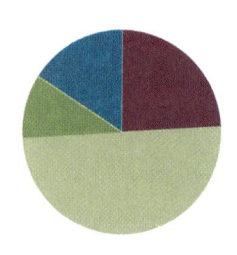

Fluency Practice (12 minutes)

- Sprint: Counting to 5 in Varied Configurations **K.CC.4b**   (12 minutes)

Sprint: Counting to 5 in Varied Configurations (12 minutes)

Materials: (S) 2 copies of the Counting to 5 Sprint

Note: In this activity, students get accustomed to the full Sprint routine while completing a task that is relatively simple conceptually. This builds confidence and enthusiasm for Sprints in the future.

Follow the instructions for delivering a Sprint in Lesson 25. Giving the identical Sprint twice facilitates comparison from Sprint A to Sprint B and allows students to see their growth. (Eventually, students will complete two Sprints that are similar but not exactly the same.) Continue to emphasize the concept of students beating their own personal score. Praise students for their hard work and for following directions in learning a new procedure.

- T: It's time for a Sprint! (Briefly recall previous Sprint preparation activities, and distribute Sprints face down.) Take out your pencil and one crayon, any color.
- T: On your mark, get set, go!
- S: (Work.)
- T: (Ring the bell, or give another signal for students to stop. Although it will not be necessary to time the students in this short practice Sprint, be sure to give the stop signal before students finish so as to not develop the expectation of finishing every time.) Pencils up!
- T: Pencils down, crayons up!
- T: It's time to check answers. What do you do if the answer is right?
- S: Circle it. (Circling correct answers instead of crossing out wrong ones avoids stigmatization.)
- T: What do you say?
- S: Yes!

A STORY OF UNITS   Lesson 28   K•3

T: (Have students correct their work, and incorporate a brief skip counting exercise including movement before Sprint B.)

T: See if you can beat your score! Race against yourself! On your mark, get set, go!

Students work on the Sprint for a second time. Perhaps give an additional three to five seconds to help students beat their first score. Give the signal to stop, reiterating that it is okay not to finish. Continue to emphasize that the goal is simply to do better than the first time. Proceed through the checking answers procedure with more enthusiasm than ever. Then, facilitate a comparison of Sprint A to Sprint B. Because students are still developing understanding of the concept of more, it may be necessary to circulate and facilitate the comparison, either visually or numerically.

T: Stand up if you beat your score.

T: You worked so hard, and I am so proud of you! Let's celebrate (e.g., congratulate each other, give three pats on the back, shake hands, have a parade).

Variation: Allow students to finish, but provide an early-finisher activity to complete on the back.

Application Problem (5 minutes)

Materials: (S) Paper, crayons, and a small ball of clay

Draw four snowmen on your paper. With your clay, make little hats and put them on the snowmen. Now, make two more hats for the snowmen that melted yesterday. How many snowmen did you draw? How many hats did you make? Which number is greater? Which number is less?

Note: This problem serves as an anticipatory set for today's lesson.

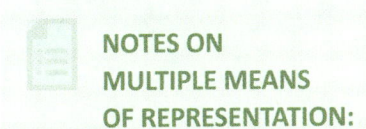

> **NOTES ON MULTIPLE MEANS OF REPRESENTATION:**
>
> Pair English language learners with a partner to facilitate the development of their understanding of the Application Problem. Teach students how to ask probing questions such as, "Do you agree?" and "Why do you think so?" as a way of extending mathematical conversations.

Concept Development (25 minutes)

Materials: (T) Bell, chime, or other gentle noisemaker (S) 1 set of 5-group cards (Template)

T: You are really good at comparing sets! I wonder if you need to see them to be able to compare them. Please close your eyes, put your heads on your desks, and listen carefully. I'm going to give you sets of sounds to compare. (Tap chime 3 times.) Think about how many chimes you just heard, and keep that number in your brain. Now, listen again. (Tap chime 6 times.) Think about the number of chimes the second time. Which number was greater?

S: 6.

T: Which number was less?

S: 3. → The first one!

T: Use your *less than* words.

S: 3 is less than 6.

Lesson 28: Visualize quantities to compare two numerals.   269

Repeat this exercise several times, using both *more than* and *less than* vocabulary, until students are confident in their answers.

> T: Now that you are confident, play a tapping game with your partner. Tap a number lightly that is less than 5. Wait. Tap another number less than 5. See if your partner can make a statement about the two numbers you tapped.

Circulate and watch students as they play. Listen for their comparison words. Allow students who are successful to work within a broader range of numbers.

> T: Next, you are going to play a game with your partner. Each of you has a mixed-up deck of number cards. Hide your deck in your hands with the number side up. When I count to three, quickly put the top card in front of you, and compare it to your partner's card. Which number is less?
>
> T: Close your eyes, and try to see how many are in each set. You may use the dots on the back to help you if you need to. When you and your partner agree, continue with the next card. (Circulate and check to ensure understanding.)

After several minutes, repeat the game. This time, however, the students should state which number is more. Circulate as they are playing to see which students still need to look at the sets in order to compare the numbers. Encourage use of *more than* and *less than* language.

Problem Set (10 minutes)

Students should do their personal best to complete the Problem Set within the allotted time.

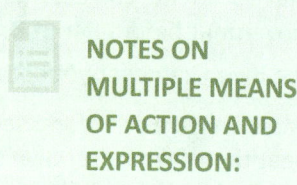

NOTES ON MULTIPLE MEANS OF ACTION AND EXPRESSION:

Model the number card game for students working below grade level. Give one instruction, model it, and ask students to demonstrate. As students show dots and numerals, vocalize the visualization "I see the number 3." "I see a number, and I see 6 dots on the back." Then, model the more than/less than statements. "3 dots is less than 6 dots." Continue observing and make note of students who still need help.

Student Debrief (8 minutes)

Lesson Objective: Visualize quantities to compare two numerals.

The Student Debrief is intended to invite reflection and active processing of the total lesson experience.

Invite students to review their solutions for the Problem Set. They should check work by comparing answers with a partner before going over answers as a class. Look for misconceptions or misunderstandings that can be addressed in the Debrief. Guide students in a conversation to debrief the Problem Set and process the lesson.

Any combination of the questions below may be used to lead the discussion.

- How did you count and compare the sets of sounds? What did you think about?
- If you are having trouble comparing two numbers, what can you do?
- When you closed your eyes, could you see a number? Who can describe how they see numbers?
- What new (or significant) math vocabulary did we use today to communicate precisely?
- How did the Application Problem connect to today's lesson?

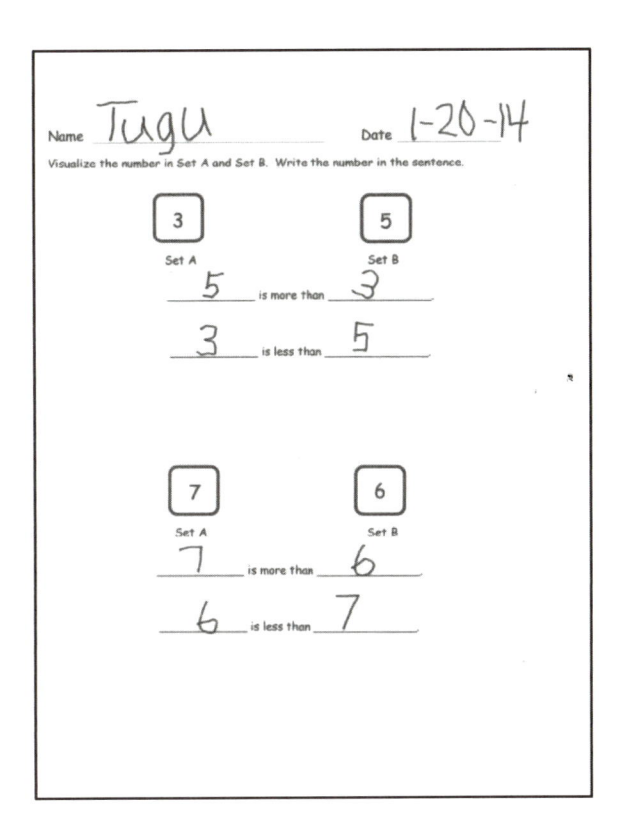

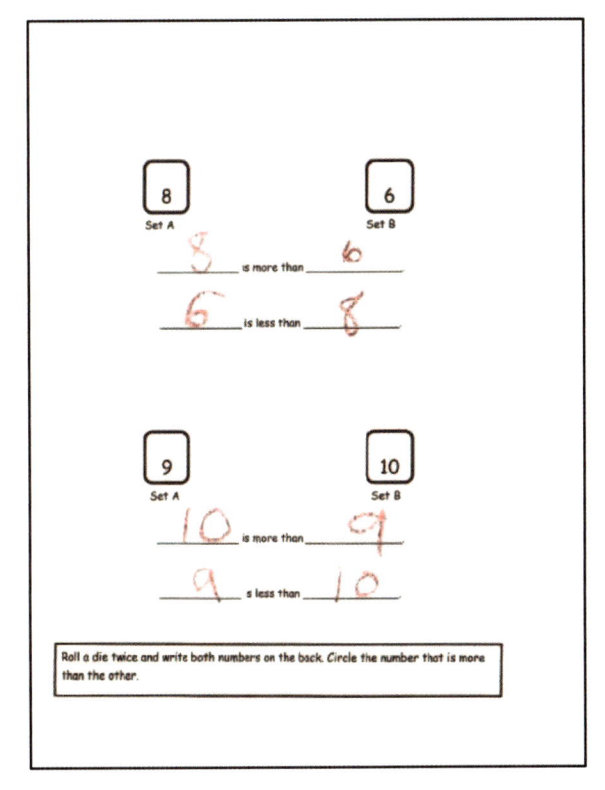

Lesson 28: Visualize quantities to compare two numerals.

Counting to 5

Lesson 28 Sprint K•3

Lesson 28: Visualize quantities to compare two numerals.

A STORY OF UNITS Lesson 28 Problem Set K•3

Name _____ Date _____

Visualize the number in Set A and Set B. Write the number in the sentences.

[3] Set A

[5] Set B

_____ is more than _____.

_____ is less than _____.

[7] Set A

[6] Set B

_____ is more than _____.

_____ is less than _____.

Lesson 28: Visualize quantities to compare two numerals. 273

A STORY OF UNITS — Lesson 28 Problem Set — K•3

[8]   [6]
Set A   Set B

_____ is more than _____.

_____ is less than _____.

[9]   [10]
Set A   Set B

_____ is more than _____.

_____ is less than _____.

Roll a die twice, and write both numbers on the back. Circle the number that is more than the other.

Lesson 28: Visualize quantities to compare two numerals.

A STORY OF UNITS   Lesson 28 Homework K•3

Name _____ Date _____

Visualize the number in Set A and Set B. Write the number in the sentences.

[7]   Set A

[4]   Set B

_____ is more than _____.

_____ is less than _____.

---

[9]   Set A

[10] Set B

_____ is more than _____.

_____ is less than _____.

Lesson 28: Visualize quantities to compare two numerals.

275

A STORY OF UNITS **Lesson 28 Homework** **K•3**

[8]
Set A

[6]
Set B

_____ is more than _____.

_____ is less than _____.

[4]
Set A

[5]
Set B

_____ is more than _____.

_____ is less than _____.

Ask a family member to give you 2 numbers. Write the numbers on the back, and circle the number that is more than the other.

276   Lesson 28:   Visualize quantities to compare two numerals.

© 2015 Great Minds. eureka-math.org
GK-M3-TE-B3-1.3.1-01.2016

EUREKA MATH

A STORY OF UNITS

Lesson 28 Template   **K•3**

0	1	2	3
4	5	5	<u>6</u>
7	8	<u>9</u>	10

5-group cards (numeral side) (Copy double-sided with 5-groups on card stock, and cut.)

Lesson 28:   Visualize quantities to compare two numerals.

A STORY OF UNITS — Lesson 28 Template — K•3

5-group cards (5-group side) (Copy double-sided with numerals on card stock, and cut.)

Lesson 28: Visualize quantities to compare two numerals.

A STORY OF UNITS

Mathematics Curriculum

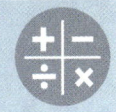

GRADE K • MODULE 3

Topic H
Clarification of Measurable Attributes

K.MD.1, K.MD.2, K.CC.6, K.CC.7

Focus Standards:	K.MD.1	Describe measurable attributes of objects, such as length or weight. Describe several measurable attributes of a single object.
	K.MD.2	Directly compare two objects with a measureable attribute in common, to see which object has "more of"/"less of" the attribute, and describe the difference. *For example, directly compare the heights of two children and describe one child as taller/shorter.*
Instructional Days:	4	
Coherence -Links from:	GPK–M4	Comparison of Length, Weight, Capacity, and Numbers to 5
-Links to:	G1–M3	Ordering and Comparing Length Measurements as Numbers

Module 3 culminates with a series of three measurement and comparison exploration tasks. In Lesson 29, students compare volume by moving a constant amount of colored water among containers of different shapes. In Lesson 30, students use balls of clay that weigh the same amount, as measured in cubes on the balance scale, to make different sculptures. They see that the same amount of clay can take various forms.

Students are challenged to plan and draw a building in Lesson 31. They compare the height of their building to that of their peers and to a linking cube stick of 10. Students then arrange their buildings to make a classroom city. When they complete the lesson, they have a new awareness of the constructions in their community.

In Module 2, students explored shapes; in Module 3, they explore the height of those shapes. For the final lesson before the End-of-Module Assessment, students consider the different measurable attributes of single items such as a water bottle, a dropper, and a juice box, as well as tools they might use to measure those attributes.

A Teaching Sequence Toward Mastery of Clarification of Measurable Attributes

Objective 1: Observe cups of colored water of equal volume poured into a variety of container shapes.
(Lesson 29)

Objective 2: Use balls of clay of equal weights to make sculptures.
(Lesson 30)

Objective 3: Use benchmarks to create and compare rectangles of different lengths to make a city.
(Lesson 31)

Objective 4: Culminating task—describe measurable attributes of single objects.
(Lesson 32)

Lesson 29

Objective: Observe cups of colored water of equal volume poured into a variety of container shapes.

Suggested Lesson Structure

- ■ Fluency Practice   (12 minutes)
- ■ Concept Development   (25 minutes)
- ■ Application Problem   (5 minutes)
- ■ Student Debrief   (8 minutes)
-   **Total Time**   **(50 minutes)**

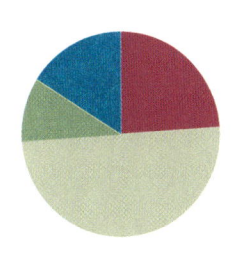

Fluency Practice (12 minutes)

- Tower Flip **K.OA.3**   (5 minutes)
- 5-Group Fill-Up **K.OA.4**   (4 minutes)
- Full, Not Full, or Empty? **K.MD.1**   (3 minutes)

Tower Flip (5 minutes)

Materials: (S) 5 linking cubes

Note: At this point in the year, many students have already mastered compositions of 3, 4, and 5. This activity seeks to build on the students' understanding of comparison to support their work with partner numbers in the next module.

- T: Touch and count your cubes.
- S: 1, 2, 3, 4, 5.
- T: How many cubes do you have?
- S: 5.
- T: Set them down on your table like a tower.
- T: Take 1 cube off the top of your tower, and place it on the table next to the tower. Do you still have 5 cubes?
- S: Yes.
- T: How many cubes are on the first tower?
- S: 4.
- T: On the other tower?
- S: 1.

T: We can say 4 and 1 make 5. Echo me, please.
S: 4 and 1 make 5.
T: Good. Take another cube off the top of the first tower, and stick it onto the top of the other tower. Do you still have 5 cubes?
S: Yes.
T: How many cubes are on the first tower?
S: 3.
T: On the other tower?
S: 2.
T: Give me the *…and…make…* statement.
S: 3 and 2 make 5.

Continue transferring cubes from one tower to the other to work through all of the combinations of 5. Then, reverse the procedure, and cycle back through the flipped combinations. Students should progress through the combinations in this order: 5 and 0, 4 and 1, 3 and 2, 2 and 3, 1 and 4, and 0 and 5. Invite students to tell what they noticed about the towers as they did this exercise (one tower got taller while the other got shorter).

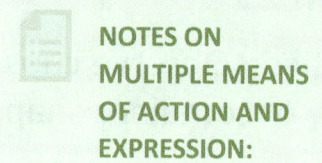

> **NOTES ON MULTIPLE MEANS OF ACTION AND EXPRESSION:**
>
> Students understand directions more quickly with a demonstration of how to compose all the combinations of 5 with linking cubes.
>
> Have students state the compositions as towers are built, and list them on the board vertically to help students see the pattern between the partner numbers.
>
> Challenge students working above grade level by having them list the combinations of 5 and state the pattern they observe between the pairs using math language.

5-Group Fill-Up (4 minutes)

Materials: (S) Dice with 6-dot side covered, personal white board

Note: This activity gives students a head start in learning how many more a number needs to make ten, anticipating the work of the next module. This activity also links to the next Fluency Practice and the numerous ways that objects can be considered full.

Conduct activity as outlined in Lesson 22.

Full, Not Full, or Empty? (3 minutes)

Materials: (T) 4 real objects filled with various amounts of liquids (e.g., small bottle, mug, vase, and bowl)

Note: A misconception students often have is that a container is full if it has any amount of liquid in it. This activity seeks to clarify the meaning of *full* in preparation for today's work with capacity.

T: Look at my water bottle. It is full because the water comes right to the top. I can't possibly put any more water in here! Repeat after me: "It is full."
S: It is full.
T: (Drink some of the water.) Now, it is not full. Echo.
S: It is not full.

A STORY OF UNITS   Lesson 29   K•3

T: (Show an empty water bottle.) This is my bottle from yesterday. There is no more water in it. Repeat after me: "It is empty."

S: It is empty.

T: Now, I'll show you some more things, and I want you to tell me if they are full, not full, or empty. (Show students a mug that is filled to the brim. Alternatively, to reduce spillage, the items could be displayed on a table or in the center of the rug with students seated on the edges of the rug so that they can see. Point to, rather than hold up, the focus object.) Raise your hand when you know what to say. (Wait for all hands to go up, and then signal.) Ready?

S: Full!

T: Very good. (Hold up a vase of flowers with a little water in it.) Raise your hand when you know what to say. (Wait for all hands to go up, and then signal.) Ready?

S: Not full!

T: Right. (Show students an empty bowl.) Ready?

S: Empty!

> **NOTES ON MULTIPLE MEANS OF ENGAGEMENT:**
>
> Students may note that one often considers a glass full even if the liquid does not come right up to the rim. Discuss reasons for that (avoiding spills, easier to drink, etc.), and let students develop their own interpretations of full based on context. The Problem Set allows room for further discussion around this topic.

Concept Development (25 minutes)

Materials: (T) Clear measuring cup, water, several vials of food coloring, an assortment of clear 1- or 2-cup capacity containers in various shapes (e.g., mug, bowl, small bottle, vase, or beaker)
(S) My capacity museum recording sheet (Template), crayons or markers

Note: Save the measuring cup and 1- or 2-cup capacity containers in various shapes for the culminating task in Lesson 32.

T: We are going to create some art today! You will be creating entries for your own Capacity Museum.

T: I have a cup of water. Student A, would you please come put two drops of red food coloring in my water container? (Assist Student A.)

T: Is my cup full?

S: Yes!

T: Watch as I pour the red water into this bowl. (Demonstrate.) Did I change the amount of water in my cup?

S: No!

T: Does it still look the same?

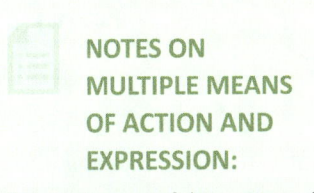

MP.7

> **NOTES ON MULTIPLE MEANS OF ACTION AND EXPRESSION:**
>
> Point to images of the concepts *full*, *not full*, and *empty* while speaking to scaffold the lesson for English language learners. For student reference, post the visuals on the word wall after introducing them.

Lesson 29: Observe cups of colored water of equal volume poured into a variety of container shapes.

283

S: No. It looks flatter. → The top of the water is wider now! → It is not as full.

T: Why do you think it looks different?

S: The bowl is bigger!

T: Yes, the bowl and the cup have different capacities. The bowl holds more water than the cup does. On your sheet, please choose one of the picture frames. Inside it, draw the bowl, and show how the water looks in the bowl.

T: I will fill my measuring cup with some new water. Student B, would you please come put two drops of blue food coloring in the cup? (Assist as necessary.)

T: I will carefully pour the blue water into this vase. (Demonstrate.) Did I change the amount of water?

S: No!

T: Does it look the same?

S: No. → Now, it looks tall. → The water is curved! → It is still full, though.

T: The cup and the vase have the same capacity but a different shape! Let's draw the water in the vase in another one of the frames on your sheet. (Continue the activity with the other colors and containers. Encourage students to notice that, while they had the same amount of water each time, it appeared to be different depending on the capacity and the shape of the container.)

Problem Set

In this lesson, the my capacity museum recording sheet serves as the Problem Set for the Concept Development.

Application Problem (5 minutes)

Demoss had a very small carton of orange juice. His mom poured it into a very tall glass without spilling any juice. Close your eyes, and think about what that might look like. Draw the little carton of juice. Now, draw the juice after she poured it into the big glass. Does Demoss have more or less juice, or does it just look different? Compare your drawings with your partner's. Are both of your glasses full? Did the glass hold all of the juice?

Note: The objective behind this Application Problem is to stimulate students' thinking about whether a change in shape necessarily results in a change in another attribute, in this case volume or capacity. Circulate during the discussion to encourage use of language such as *more than, less than*, and *the same as*. Note from the drawings which students might need extra support understanding this concept.

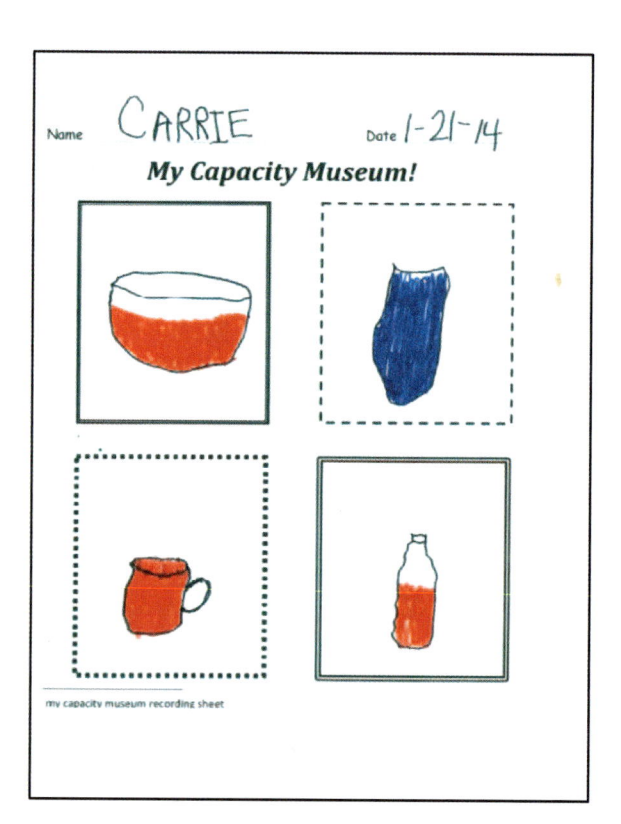

Student Debrief (8 minutes)

Lesson Objective: Observe cups of colored water of equal volume poured into a variety of container shapes.

The Student Debrief is intended to invite reflection and active processing of the total lesson experience.

Invite students to review their recording sheets. They should compare answers with a partner before reviewing observations as a class. Look for misconceptions or misunderstandings that can be addressed in the Debrief.

Any combination of the questions below may be used to lead the discussion.

- Why did the water look different in each of the containers?
- Did the amount of the water change each time?
- Turn to your partner, and compare your drawings. Do they look the same?
- Which container do you think would hold the most?
- How did you determine if a container was empty, not full, or full?
- How did you know when a container was full?
- How can *full* be different in certain situations? (For example, with a mug of hot chocolate, you don't want to fill it too full and spill.)
- When do we need *full* to mean right to the top?
- What new (or significant) math vocabulary did we use today to communicate precisely?
- How did the Application Problem connect to today's lesson?

A STORY OF UNITS   Lesson 29 Homework   K•3

Name _____   Date _____

Draw a line from each container to the word that describes the amount of liquid the container is holding.

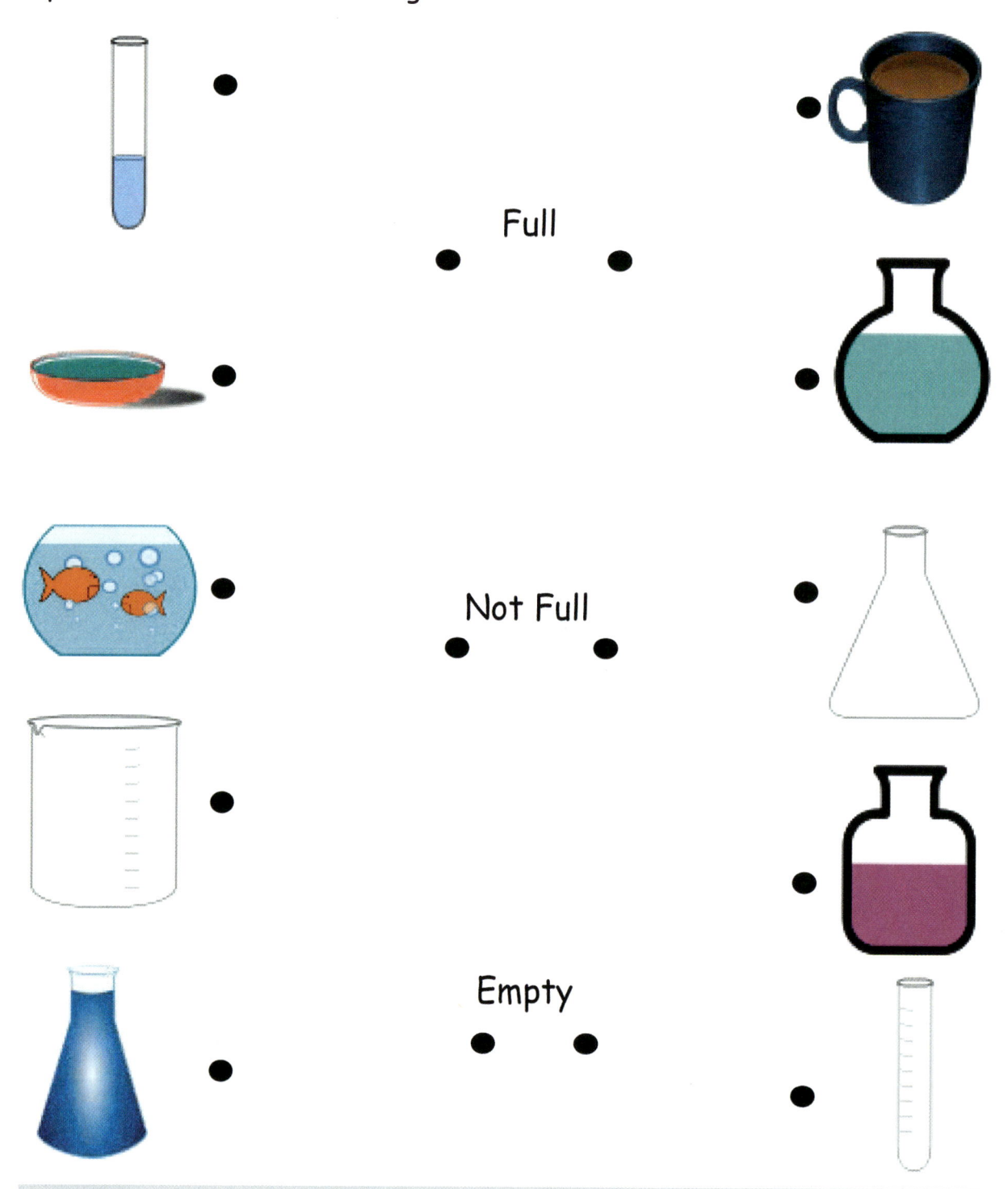

Full

Not Full

Empty

286   Lesson 29:   Observe cups of colored water of equal volume poured into a variety of container shapes.

A STORY OF UNITS

Lesson 29 Template K•3

Name _____ Date _____

My Capacity Museum!

my capacity museum recording sheet

Lesson 29: Observe cups of colored water of equal volume poured into a variety of container shapes.

| A STORY OF UNITS | Lesson 30 K•3 |

Lesson 30

Objective: Use balls of clay of equal weights to make sculptures.

Suggested Lesson Structure

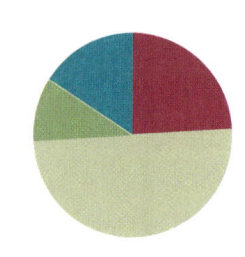

- ■ Fluency Practice   (12 minutes)
- ■ Application Problem   (5 minutes)
- ■ Concept Development   (25 minutes)
- ■ Student Debrief   (8 minutes)
-   **Total Time**   **(50 minutes)**

Fluency Practice (12 minutes)

- Tower Flip **K.OA.3**   (4 minutes)
- Counting the Say Ten Way with the Rekenrek **K.NBT.1**   (3 minutes)
- Growing Apples to 10 **K.OA.4**   (5 minutes)

Tower Flip (4 minutes)

Materials: (S) 5 linking cubes

Note: In this activity, students see that the relationship between the quantities remains the same even though the orientation has changed from the previous iteration, from height to length.

Conduct the activity as outlined in Lesson 29, but this time, have students lay the towers down on the table and refer to them as trains. While transferring cubes from one to the other, guide students to notice that as one train becomes longer, the other becomes shorter.

Counting the Say Ten Way with the Rekenrek (3 minutes)

Materials: (T) 20-bead Rekenrek

Note: This activity is an extension of students' previous work with the Rekenrek and anticipates working with teen numbers.

Conduct the activity as described in Lesson 6, but continue to 2 tens, or 10 and 10.

Growing Apples to 10 (5 minutes)

Materials: (S) Apple mat (Fluency Template), 10 red beans, die with 6-dot side covered

Note: This activity gives students a head start in learning partners to ten, anticipating the work of the next module.

1. Roll the die.
2. Put that many red beans on the apple tree, arranging them in 5-groups.
3. Count how many more are needed to make 10.
4. Say, "I have ____. I need _____ more to make 10."
5. Do not remove the beans. Roll the die again. Count to see if there are enough spaces for that many beans. (If the number goes over 10, and there aren't enough spaces, simply roll again to get a smaller number.) Then, place that many beans on the apple tree.
6. State the new amount and how many more it needs to make 10.

Continue until 10 is made. Then, remove the beans, and start again from 0 if time permits. This game can also be played with a partner. A spinner can be used instead of a die.

Application Problem (5 minutes)

Imagine a balance scale. Can you see it? Now, imagine putting one big ball of clay on one side and four small balls of clay on the other. If the big ball is as heavy as the four small balls put together, then what would the balance scale look like? Draw it.

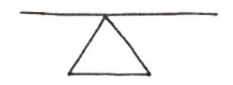

Note: In this problem, we turn the thinking around to ask the students to consider and demonstrate what they know about the conservation of weight. Circulate during the discussion to ensure understanding of the question and to see if students are creating reasonably sized balls for the comparison. Encourage correct mathematical vocabulary.

> **NOTES ON MULTIPLE MEANS OF REPRESENTATION:**
>
> Give English language learners and students working below grade level a graphic that includes a picture of a balance scale to help them understand what is being asked. Remind them how the balance scale looks when two objects are the same weight. The visual helps students demonstrate that the balance is even or level. Check students' learning by asking them to explain what it means when the balance scale is balanced.

> **NOTES ON MULTIPLE MEANS OF ENGAGEMENT:**
>
> Ask students working above grade level to explain why the differently shaped objects made with the clay still weigh the same when placed on the balance scale. Ask them how they would go about proving their thinking to an alien who just landed on earth.

Lesson 30: Use balls of clay of equal weights to make sculptures.

Concept Development (25 minutes)

Materials: (S) Balance scale, 2 small pieces of clay per pair of partners (different color clays, but equal weight), clay shapes recording sheet (Template)

Note: Save a piece of clay for the culminating task in Lesson 32.

- T: Work with your partner. Test your pieces of clay on your balance scale. What do you notice?
- S: They are the same weight. → It balances!
- T: With all of your blue clay, make a snake as long as your pinky finger. With all of your red clay, make a snake as long as your thumb. What do you notice?
- S: The red snake is longer than the blue. → The blue snake is fatter.
- T: Test your snakes on the balance.
- S: They weigh the same!
- T: What is the same about them?
- S: They're the same weight.
- T: Now, make a little bowl with your blue clay. On your Recording Sheet, pick a frame. In that frame, draw a picture of your bowl. Make a cup with your red clay. Draw a picture of your cup in another frame. Now, test them on your balance.
- S: It still balances. → This one looks bigger, but it still weighs the same as the other one.
- T: Hmm. I wonder which of your clay containers would hold more? (Allow time for discussion and sharing.)

Continue the lesson, experimenting and recording results with other shapes and configurations: flatter and taller, wider and narrower, thinner and fatter, and so on. Lead the children to discover that the weight is constant regardless of the configuration of the clay.

- T: Now, make your favorite indoor animal with the blue clay, and draw it on your Recording Sheet. Make your favorite outdoor animal with the red clay, and draw it on your Recording Sheet. Can you make a guess about their weights? Talk about your guess with your partner. Now, test your guess!

Problem Set

In this lesson, the clay shapes recording sheet serves as the Problem Set for the Concept Development.

Student Debrief (8 minutes)

Lesson Objective: Use balls of clay of equal weights to make sculptures.

The Student Debrief is intended to invite reflection and active processing of the total lesson experience.

Invite students to compare their recording sheets with a partner before reviewing as a class. Look for misconceptions or misunderstandings that can be addressed in the Debrief.

Any combination of the questions below may be used to lead the discussion.

- When you compared your clay with your partner's, did you expect them to weigh the same? Why or why not?
- What did you learn when you and your partner made snakes and compared their weights? Were you surprised with the results? Did this change your thinking? (Observe student responses to see who has a grasp of the conservation of weight.)
- What happened to the weight of your clay when you created a new object? Was this a surprise to you, or did you expect your clay to weigh the same?
- How did the Application Problem connect to today's lesson?

Lesson 30 Homework K•3

Dear Parents:

In class, we used balls of clay that weigh the same on the balance scale to make different sculptures. We saw that the same balls of clay can take various forms without changing the weight. The balls weighed the same on the balance scale, as did the sculptures.

Today's homework is a review of fluency work.

A STORY OF UNITS   Lesson 30 Homework   K•3

Name _____   Date _____

Color 4 apples.   Color 2 apples.   Color 7 apples.

I colored _____ apples.   I colored _____ apples.   I colored _____ apples.

I need to color _____ more to I need to color _____ more to I need to color _____ more to
make 10.   make 10.   make 10.

Color 1 apple.   Color 9 apples.   Color 3 apples.

I colored _____ apples.   I colored _____ apples.   I colored _____ apples.

I need to color _____ more to I need to color _____ more to I need to color _____ more to
make 10.   make 10.   make 10.

Lesson 30:   Use balls of clay of equal weights to make sculptures.

A STORY OF UNITS

Lesson 30 Fluency Template   K•3

apple mat

294 Lesson 30: Use balls of clay of equal weights to make sculptures.

© 2015 Great Minds. eureka-math.org
GK-M3-TE-B3-1.3.1-01.2016

EUREKA MATH

Name _____   Date _____

Clay Shapes

clay shapes recording sheet

Lesson 30: Use balls of clay of equal weights to make sculptures.

Lesson 31

Objective: Use benchmarks to create and compare rectangles of different lengths to make a city.

Suggested Lesson Structure

- Fluency Practice   (12 minutes)
- Application Problem (5 minutes)
- Concept Development (25 minutes)
- Student Debrief   (8 minutes)

Total Time   **(50 minutes)**

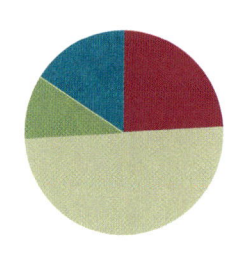

Fluency Practice (12 minutes)

- Sprint: Rekenrek to 5 **K.CC.5**   (12 minutes)

Sprint: Rekenrek to 5 (12 minutes)

Materials: (S) 2 copies of the Rekenrek to 5 Sprint

Note: In this activity, students grow more comfortable with the Sprint routine while completing a task that involves relatively simple concepts. This builds confidence and enthusiasm for Sprints.

Follow the instructions for delivering a Sprint in Lesson 25. Use the Rekenrek to 5 Sprint for both rounds. Giving the identical Sprint twice facilitates comparison from Sprint A to Sprint B and allows students to see their growth. (Eventually, students will complete two Sprints that are similar but not exactly the same.) Continue to emphasize the concept of students beating their personal score. Praise students for their hard work and for following directions.

Application Problem (5 minutes)

Materials: (S) Bag of pony beads, 1 foot of elastic string or yarn with a bead tied on one end to prevent the beads from falling off

Using your elastic or your yarn, make a string of beads that is as long as your hand. Turn to your partner to talk about how you decided how long to make your string. Compare your strings.

Are they the same length? Tie the ends of your string together to make a bracelet!

> **NOTES ON MULTIPLE MEANS OF REPRESENTATION:**
>
> Model the Application Problem for students working below grade level and English language learners. Help them compare the lengths of their bracelets by modeling what to say: "My bracelet is longer than/shorter than yours because...."

| A STORY OF UNITS | Lesson 31 | K•3 |

Note: This problem serves as an anticipatory set for the exercise of creating something as long as a chosen benchmark. Circulate during the activity to determine if children need support determining where their hand "ends."

Concept Development (25 minutes)

Materials: (S) Construction paper, crayons or markers, scissors, tape, 10-sticks

MP.6

T: Today we are going to make a math city! We will use construction paper for each of you to design a special building for our city. First, plan how tall you want your building to be. Think about comparing the height of your building to something else in the room. What are some of your ideas?

S: I'm going to make my building taller than a 10-stick. → I want to make a house shorter than my hand. → I'm going to make a skyscraper as long as my foot! → My building will be just as long as my pencil.

T: Now, you need to think about the shape and color of your building. Turn to your partner, and talk about your plan. What type of building do you want it to be?

Allow time for sharing and discussion.

T: You may begin your work. I will be visiting all of you during your work to see how you thought about the height of your building. What are you comparing it to? I will help you write your answer on the back.

Allow time for students to work. Circulate during the work period, and ask students about the height of their buildings. Ask them to show you how they compared the height of their buildings to specific classroom objects. Write their answers on the back.

> **NOTES ON MULTIPLE MEANS OF ENGAGEMENT:**
>
> Challenge students working above grade level by asking them to think about and explain how comparing their buildings to something else in the room helps the class create the math city. Encourage them to use their math vocabulary in expressing their ideas.

T: Now we will create our city! Students A and B, please bring your buildings to the front. Whose is shorter?

S: Student A's.

T: Great! Please find a place on the bulletin board for your buildings. (Help students affix their work to the wall or bulletin board.)

T: Students C and D, please bring up your buildings. Whose is taller?

S: Student C's.

T: Good! Please find a place in the city for your buildings. (Continue with sets of student work, each time comparing the heights of the buildings and reinforcing *taller than* and *shorter than* language.)

T: This is a wonderful city! Take some time to talk about the city with your friends. Which buildings do you think would be taller than your foot? Which ones do you think would be shorter than your hand? Are there any that would be shorter than a crayon? (Allow time for observation and discussion. Encourage students to use benchmarks for their comparison: "Here is my pencil! This building is longer, but this one is shorter than my pencil!")

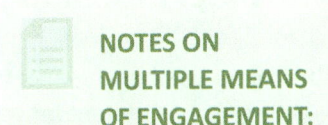

Lesson 31: Use benchmarks to create and compare rectangles of different lengths to make a city.

Problem Set (10 minutes)

Materials: (S) 5-stick

Students should do their personal best to complete the Problem Set within the allotted time.

Read the directions below, and have students draw the imaginary animal inside the box.

1. Draw a rectangle body as long as a 5-stick.
2. Draw 4 rectangle legs each as long as your thumb.
3. Draw a circle for a head as wide as your pinky.
4. Draw a line for a tail shorter than your pencil.
5. Draw in eyes, a nose, and a mouth.

Student Debrief (8 minutes)

Lesson Objective: Use benchmarks to create and compare rectangles of different lengths to make a city.

The Student Debrief is intended to invite reflection and active processing of the total lesson experience.

Invite students to review their solutions for the Problem Set. They should check work by comparing answers with a partner before going over answers as a class. Look for misconceptions or misunderstandings that can be addressed in the Debrief. Guide students in a conversation to debrief the Problem Set and process the lesson.

Any combination of the questions below may be used to lead the discussion.

- How did you choose how tall you wanted your building to be?
- How did you choose the object to compare your building to?
- Did you test to see if your guess was right?
- Compare your imaginary animal to a partner's. Do they look the same? How are they different?
- Why would your drawings be different if you followed the same directions? Were your comparisons different?
- What new (or significant) math vocabulary did we use today to communicate precisely?
- How did the Application Problem connect to today's lesson?

A STORY OF UNITS

Lesson 31 Sprint K•3

Rekenrek to 5

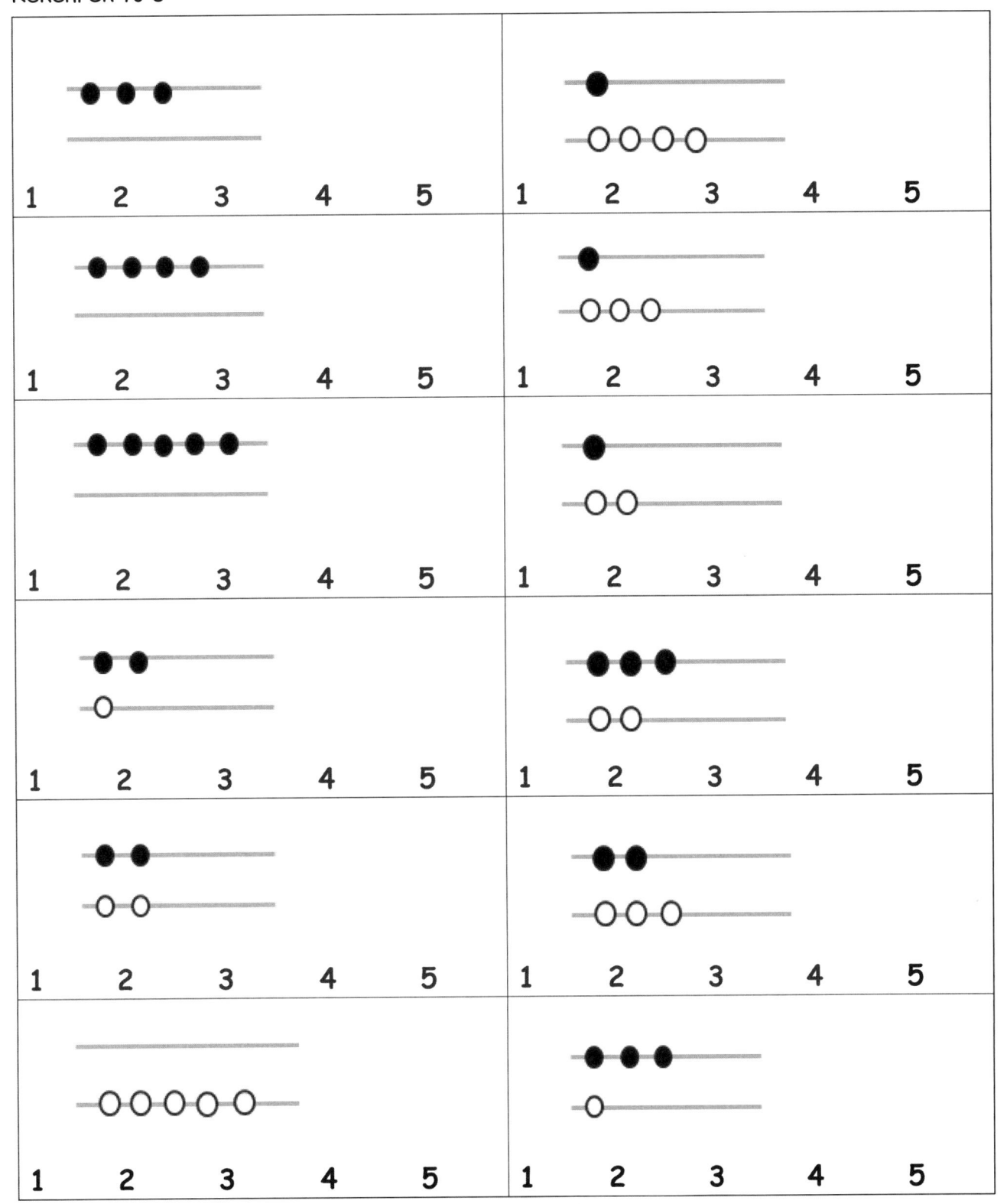

Lesson 31: Use benchmarks to create and compare rectangles of different lengths to make a city.

Name _____ Date _____

Listen to the directions, and draw the imaginary animal inside the box.

Draw a rectangle body as long as a 5-stick.
Draw 4 rectangle legs each as long as your thumb.
Draw a circle for a head as wide as your pinky.
Draw a line for a tail shorter than your pencil.
Draw in eyes, a nose, and a mouth.

Imaginary Animal

Name _____ Date _____

Read the following directions to your child to make a castle:
- Draw a rectangle as long as a spoon.
- Draw another rectangle on each side of the rectangle you just made.
- Draw a triangle on top of each rectangle to make towers shorter than your hand.
- Draw 1 rectangle flag pole as long as your pointer finger.
- Draw 1 square flag as long as your pinky.
- Draw a door as long as your thumb.
- Draw 2 hexagon windows each as long as a fingernail.
- Draw a prince or princess in your castle.

Castle

Lesson 31: Use benchmarks to create and compare rectangles of different lengths to make a city.

Lesson 32

Objective: Culminating task—describe measurable attributes of single objects.

Suggested Lesson Structure

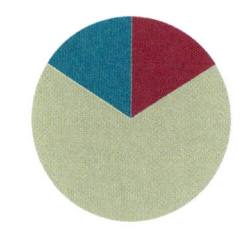

- **Fluency Practice** (8 minutes)
- **Concept Development** (34 minutes)
- **Student Debrief** (8 minutes)
- **Total Time** **(50 minutes)**

Fluency Practice (8 minutes)

- Breaking Apart Dot Cards of 6 **K.CC.2**   (4 minutes)
- Mystery Attribute **K.MD.2**   (4 minutes)

Breaking Apart Dot Cards of 6 (4 minutes)

Materials: (S) Dot cards of 6 (Lesson 13 Fluency Template) inserted into personal white board

Note: Students decompose numbers pictorially in this activity and develop part–whole thinking, essential to the work of the next module.

1. Have students touch and count the dots.
2. Partner A circles a group of dots, and then tells how many he circled.
3. Partner B tells how many are not circled and gives a (*and … make*) statement (e.g., 4 and 2 make 6).
4. Partners erase, switch roles, and continue exploring compositions of 6.

Variation: Give two (*and … make*) statements when applicable (e.g., 4 and 2 is 6; 2 and 4 is 6), or give two *6 is…* statements when applicable (e.g., 6 is 5 and 1; 6 is 1 and 5).

Mystery Attribute (4 minutes)

Materials: (T) Assorted classroom objects, balance scale

Note: This activity challenges students by presenting multiple attributes, preparing students for the culminating task.

- T: (Show students a pencil and crayon side by side, vertically, with endpoints aligned.) Listen carefully, and raise your hand when you know what word is missing: "The pencil is _____ than the crayon." (If students are unsure at first, offer two options—taller or heavier.) Ready?
- S: Taller! → Longer!

Repeat with pencil and crayon side by side, horizontally, with endpoints aligned. Repeat with objects on a balance scale. Continue with a variety of objects, having students identify the attribute by indicating taller or shorter, longer or shorter, or heavier or lighter.

| A STORY OF UNITS | Lesson 32 | K•3 |

Application Problem

Note: In this lesson, the Application Problem has been omitted to allow more time for the culminating task.

Concept Development (34 minutes)

Materials: (T) Wide variety of objects arranged on the table from past lessons such as a piece of clay, a few linking cube sticks, clear containers including a vase and a cup, a string, a paper strip, a set of pennies, an empty clean juice box, a water bottle, and other student favorites
(S) Balance scale, bag of two cups of rice, small scoop, and tray for a working surface per pair, comparing attributes recording sheet (Template)

T: You have learned so much about how to compare things! We are going to play a comparing game today. Student A, please come up to the table. Choose an object.

S: I chose the cup.

T: If you wanted to tell someone about the cup, what would you say?

S: It is clear. It is about as high as my finger. It is not heavy! It could hold as much rice as I could hold in my hands!

T: We can talk about the cup in a lot of different ways, can't we? We can talk about its height, its weight, or its capacity. (Act out each as you use the words.) These are all ways to describe and compare objects.

T: Student B, please come up and choose two objects on the table.

S: (Answers may vary.) I chose the linking cube stick and the vase.

T: What are some things that are different about the stick and the vase?

S: The vase can hold things, but the stick is taller! I think the vase feels heavier, too.

T: Let's choose one way to compare first. Let's compare which is heavier. Use the balance scale to see which is heavier. (Allow the student to demonstrate.) Which is heavier? Use your math words!

S: The vase is heavier than the cube stick.

T: Yes, the cube stick is lighter than the vase. Let's show how we would put that on our Recording Sheet. (Demonstrate by drawing a balance and the objects.)

T: Let's think about your other ideas. Which do you think is taller? How could you find out?

S: We have to line them up first. The stick is taller than the vase!

T: Yes, the vase is shorter than the stick. The stick is taller than the vase. Let's draw that on our Recording Sheet. (Demonstrate.)

T: What other way could we compare the vase and the cube stick?

S: We could see which holds more. → We could find its capacity!

T: Yes, we have been talking about the capacity of things. We were thinking about which object can hold more. What do you think about the vase and the cube stick?

S: The vase could hold some rice, but the cube stick can't really hold anything.

T: Okay, so I can figure out the capacity of the vase, but is the cube stick meant to hold liquid or objects?

S: No.

Lesson 32: Culminating task—describe measurable attributes of single objects.

A STORY OF UNITS   **Lesson 32**   K•3

MP.5

T: The cube stick might be able to hold a little bit of liquid or small objects in these dents, but we don't actually use it for that. So, does it make sense to compare the capacity of these two objects?

S: No!

T: We can compare things in lots of different ways! Our answers will be very different depending on what we choose to compare and how we compare them.

T: You will work with a partner. You will choose a pair of objects and decide in what ways you could compare them. Talk to your partner about which way is the best way to measure. You will use the recording sheet to draw a picture to show which is more. Think about your math words. Is one object longer than the other? Does it hold more than the other? Is it heavier than the other? Test your guesses, and show your work! When you have recorded your work, choose two other objects.

Allow ample time for discussion and experimentation. Circulate to ensure isolation of individual attributes and correct comparisons on the recording sheet. Encourage correct math vocabulary.

T: Let's share some of our discoveries! Who would like to share?

Student Debrief (8 minutes)

Lesson Objective: Culminating task—describe measurable attributes of single objects.

The Student Debrief is intended to invite reflection and active processing of the total lesson experience.

Invite students to review their Recording Sheet. They should compare work with a partner before reviewing it with the class. Look for misconceptions or misunderstandings that can be addressed in the Debrief.

Any combination of the questions below may be used to lead the discussion.

- How did you and your partner decide to compare your first set of objects?
- What did you discover?
- How did you draw your discovery on your recording sheet to show your friends?
- Are there any objects that you couldn't compare in a certain way? Why?
- What new (or significant) math vocabulary did we use today to communicate precisely?

Lesson 32:   Culminating task—describe measurable attributes of single objects.

© 2015 Great Minds. eureka-math.org
GK-M3-TE-B3-1.3.1-01.2016

EUREKA MATH

A STORY OF UNITS — Lesson 32 Homework K•3

Name _____ Date _____

The homework is a review of fluency skills from Module 3.

Circle a group of dots. Then, fill in the blanks to make a number sentence.

_2_ and _4_ is _6_

____ and ____ is ____.

____ and ____ is ____.

____ and ____ is ____.

____ and ____ is ____.

On the back, make your own 6-dot cards. Circle some dots, and then say, "____ and ____ is ____."

Lesson 32: Culminating task—describe measurable attributes of single objects.

Name _____   Date _____

comparing attributes recording sheet

A STORY OF UNITS

End-of-Module Assessment Task K•3

Student Name _____

Topic E: Are There Enough?

Rubric Score: _____ Time Elapsed: _____

	Date 1	Date 2	Date 3
Topic E			
Topic F			
Topic G			
Topic H			

Materials: (T) 7 spoons, 8 bowls, 6 1 inch × 1 inch squares, 1 2 inch × 3 inch square piece of paper

1. Is there enough space on this paper for all these squares? Show me how you know.
2. Are there enough spoons for the bowls? Show me how you know.
3. Use the words *more than* to compare the spoons and bowls.
4. Use the words *less than* to compare the spoons and bowls.

What did the student do?	What did the student say?
1.	
2.	
3.	
4.	

EUREKA MATH

Module 3: Comparison of Length, Weight, Capacity, and Numbers to 10

© 2015 Great Minds. eureka-math.org
GK-M3-TE-B3-1.3.1-01.2016

Topic F: Comparison of Sets Within 10

Rubric Score: _____ Time Elapsed: _____

Materials: (S) 1 set of 6 linking cubes, 1 set of 4 linking cubes, additional linking cubes

1. Which set has more cubes? (Show the set of 6 cubes and the set of 4 cubes.)
2. Make a set that has the same number of cubes as this one. (Present the set with 4 cubes.) Tell me what you are doing.
3. Make a set that has 1 more cube than this set. (Present the set with 6 cubes.)
4. Make a set that has 1 less cube than this set. (Present a set with 10 cubes.)

What did the student do?	What did the student say?
1.	
2.	
3.	
4.	

Topic G: Comparison of Numerals

Rubric Score: _____ Time Elapsed: _____

Materials: (T) 12 loose linking cubes

1. (Present a set with 7 cubes and a set with 5 cubes.) Put these objects in lines to match and compare them.
2. Which number is more? Less?
3. (Write the numerals 8 and 4.) Use the words *more than* to compare these two numerals.

What did the student do?	What did the student say?
1.	
2.	
3.	

Module 3: Comparison of Length, Weight, Capacity, and Numbers to 10

A STORY OF UNITS

End-of-Module Assessment Task K•3

Topic H: Clarification of Measurable Attributes

Rubric Score: _____ Time Elapsed: _____

Materials: (T) Empty juice box with the top cut off, cup full of rice, linking cube stick of 7, balance scale, many additional cubes, student scissors, tub for pouring rice from cup to juice box

1. Compare the length of this juice box to the length of this stick. Use your words.
2. Compare the weight of this juice box to the weight of this pair of scissors. Use your words.
3. Compare the weight of this juice box to the weight of the cubes. How many cubes weigh the same as the juice box? Use your words. (If the student does not use the balance scale but makes a thoughtful guess, encourage use of the scale to confirm the estimate.)
4. Compare the capacity of this juice box to this cup.

What did the student do?	What did the student say?
1.	
2.	
3.	
4.	

Module 3: Comparison of Length, Weight, Capacity, and Numbers to 10

| A STORY OF UNITS | End-of-Module Assessment Task | K•3 |

| End-of-Module Assessment Task Standards Addressed | Topics E–H |

Compare numbers.

K.CC.6 Identify whether the number of objects in one group is greater than, less than, or equal to the number of objects in another group, e.g., by using matching and counting strategies. (Include groups with up to ten objects.)

K.CC.7 Compare two numbers between 1 and 10 presented as written numerals.

Describe and compare measurable attributes.

K.MD.1 Describe measurable attributes of objects, such as length or weight. Describe several measurable attributes of a single object.

K.MD.2 Directly compare two objects with a measurable attribute in common, to see which object has "more of"/"less of" the attribute, and describe the difference. *For example, directly compare the heights of two children and describe one child as taller/shorter.*

Evaluating Student Learning Outcomes

A Progression Toward Mastery is provided to describe and quantify steps that illuminate the gradually increasing understandings that students develop *on their way to proficiency*. In this chart, this progress is presented from left (Step 1) to right (Step 4). The learning goal for students is to achieve Step 4 mastery. These steps are meant to help teachers and students identify and celebrate what the students CAN do now and what they need to work on next.

A STORY OF UNITS

End-of-Module Assessment Task   K•3

A Progression Toward Mastery				
Assessment Task Item and Standards Assessed	STEP 1 Little evidence of reasoning without a correct answer. (1 Point)	STEP 2 Evidence of some reasoning without a correct answer. (2 Points)	STEP 3 Evidence of some reasoning with a correct answer or evidence of solid reasoning with an incorrect answer. (3 Points)	STEP 4 Evidence of solid reasoning with a correct answer. (4 Points)
Topic E K.CC.6	Student is largely unresponsive and unable to perform the tasks.	Student shows evidence of beginning to understand comparison but makes many errors.	Student is able to complete the tasks but may be unable to use his words correctly in the third and fourth questions.	Student correctly: • Places the squares on the paper to see if they fit. • Shows there are not enough spoons for the bowls. • Uses the words *more than* and *less than* to compare the spoons and bowls.
Topic F K.CC.6	Student is largely unresponsive and unable to perform the tasks.	Student demonstrates a beginning understanding of comparison but makes many small errors.	Student demonstrates understanding of comparison but makes a small error, for example: • Unable to state that 6 is more than 4. • Struggles with showing one of the following sets: 1 more than 6, 1 less than 10, or a set equal to 4.	Student correctly: • Shows which set is more and states that 6 is more than 4. • Shows a set equal to 4. • Shows a set 1 more than 6. • Shows a set 1 less than 10.

312   Module 3:   Comparison of Length, Weight, Capacity, and Numbers to 10

EUREKA MATH

A Story of Units — End-of-Module Assessment Task — K•3

A Progression Toward Mastery

Topic G **K.CC.6** **K.CC.7**	Student shows little evidence of comparison and is unable to articulate thoughts.	Student shows evidence of beginning to understand comparison but has not yet mastered the language of comparison.	Student makes a small error such as: • Omitting the word *than* when using comparison words or confuses *less than* with *more than*, though knows which number is larger and more than, even though it is evident.	Student correctly: • Puts the objects in lines to match and compare them. • Uses *more than* and *less than* to compare 7 and 5. • Compares the numerals 8 and 4.
Topic H **K.MD.1** **K.MD.2**	Student shows little evidence of understanding what is being asked.	Student shows evidence of beginning to understand comparison but has not yet mastered the language of comparison.	Student makes one error, such as: • Confuses measurement vocabulary or does not use tools but makes intelligent surmises backed by reasoning.	Student correctly uses language and the appropriate tools to: • Compare the length of the box to the stick. • Compare the weight of the box to the scissors. • Compare the weight of the box to a number of cubes on the balance scale. • Compare the capacity of the box using the rice.

End-of-Module Assessment Task

A STORY OF UNITS — K•3

Class Record Sheet of Rubric Scores: Module 3

Name:	Topic E: Are There Enough?	Topic F: Comparison of Sets Within 10	Topic G: Comparison of Numerals	Topic H: Clarification of Measurable Attributes	Next Steps:

Module 3: Comparison of Length, Weight, Capacity, and Numbers to 10

Answer Key

Eureka Math® Grade K Module 3

Special thanks go to the Gordon A. Cain Center and to the Department of Mathematics at Louisiana State University for their support in the development of *Eureka Math*.

A STORY OF UNITS

Mathematics Curriculum

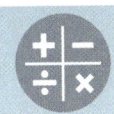

K GRADE

GRADE K • MODULE 3

Answer Key
GRADE K • MODULE 3

Comparison of Length, Weight, Capacity, and Numbers to 10

Lesson 1

Problem Set

First tower circled; second tower circled; second tower circled

Flower drawn taller than vase; tree drawn taller than house

Glue circled; second hammer circled; first lightning bolt circled

Bookmark drawn shorter than book; crayon drawn shorter than pencil

Homework

3 shorter flowers drawn; 5

2 taller ladybugs drawn; 4

Object drawn taller than student; object drawn shorter than flag pole

Lesson 2

Recording Sheet

Answers will vary.

Problem Set

Pants, car, key, and left side of door colored orange

Glove, leaf, butterfly, and top of door colored green

Answers will vary.

Homework

Answers will vary.

Lesson 3

Problem Set

Answers will vary.

Homework

Vehicles shorter than crayon circled in blue

Vehicles longer than crayon circled in red

Answers will vary.

Lesson 4

Problem Set

4-stick circled, 4; 3-stick circled, 3

6-stick circled, 6; 7-stick circled, 7

Stick shorter than 5 drawn

Stick longer than 6 drawn

Stick shorter than 9 drawn

Homework

1-, 2-, 3-, and 4-sticks circled in red

6-, 7-, and 9-sticks circled in blue

7-stick drawn; 1 stick longer than 7-stick drawn; 1 stick shorter than 7-stick drawn

Lesson 5

Problem Set

2-stick circled; 3-stick circled

5-stick circled

5, 4; 4, 5

7-stick circled

9, 7; 7, 9

6-stick drawn; 1 stick longer than 6-stick drawn; 1 stick shorter than 6-stick drawn
OR
9-stick drawn; 1 stick longer than 9-stick drawn; 1 stick shorter than 9-stick drawn

Homework

6-stick circled

6, 7; 7, 6

7-stick drawn; 1 stick longer than 7-stick drawn; 1 stick shorter than 7-stick drawn

9-stick circled

8, 9; 9, 8

5-stick drawn; 1 stick longer than 5-stick drawn; 1 stick shorter than 5-stick drawn

Lesson 6

Problem Set

Stick circled blue; 6

Stick circled green; 9

3-stick traced; crayon traced

Stick longer than marker traced; marker traced

5-stick traced; object longer than 5-stick traced

Homework

7 cubes next to palm tree colored

4 cubes under crayon colored

3 cubes under bus colored

4 cubes under shoe colored

Lesson 7

Problem Set

2 cubes colored red; 3 cubes colored green; 5; YES circled

1 cube colored red; 4 cubes colored green; 5; YES circled; 5

6-stick traced; object the same length drawn

7-stick traced; object the same length drawn

8-stick traced; object the same length drawn

Homework

2 cubes colored green; 3 cubes colored blue

3 cubes colored blue; 2 cubes colored green; 5

1 cube colored green; 4 cubes colored blue; 5

4 cubes colored green; 1 cube colored blue; 5

2 cubes colored yellow; 2 cubes colored blue; 4

Lesson 8

Fluency

1 oval drawn; 5 ovals circled

4 triangles drawn; 5 triangles circled

2 squares drawn; 5 squares circled

1 oval drawn; 5 ovals circled

1 triangle drawn; 5 triangles circled

1 square drawn; 5 squares circled

1 oval crossed out; 5 ovals circled

1 triangle crossed out; 5 triangles circled

5 squares circled

2 triangles crossed out; 5 triangles circled

2 squares crossed out; 5 squares circled

2 ovals crossed out; 5 ovals circled

1 triangle crossed out; 5 triangles circled

1 square crossed out; 5 squares circled

3 ovals crossed out; 5 ovals circled

3 triangles crossed out; 5 triangles circled

3 squares crossed out; 5 squares circled

7 ovals crossed out; 5 ovals circled

Problem Set

Book circled; pen circled

Large bear circled; shoe circled

Tennis ball circled; watermelon circled

Answers will vary.

Homework

Answers will vary.

Lesson 9

Recording Sheet

Answers may vary.

Homework

Answers will vary.

Lesson 10

Recording Sheet

Answers may vary.

Homework

6; 9

5 pennies drawn; 10 pennies drawn

Answers will vary.

Lesson 11

Problem Set

Line drawn from 3 cubes to 3 cubes

Line drawn from 4 cubes to 4 cubes

Line drawn from 6 cubes to 6 cubes

Line drawn from 10 cubes to 10 cubes

Homework

1 cube drawn; 1 cube drawn

1 cube drawn; 1 cube drawn

9 cubes drawn

Lesson 12

Recording Sheet

Answers will vary.

Homework

8; 2

10; 6

Lesson 13

Recording Sheet

Answers may vary.

Problem Set

Verbal answers may vary.

Homework

2 sets within each set of 6 circled; answers may vary.

Lesson 14

Recording Sheet

Answers may vary.

Homework

1 set of 6 circled for each; answers may vary.

Lesson 15

Recording Sheet

Answers may vary.

Homework

2 sets within each set of 7 circled

Lesson 16

Recording Sheet

Answers may vary.

Problem Set

10 squares cover the shape; 10

Answers may vary.

Homework

Student's hand traced; adult's hand traced

Bigger hand size recorded; answers may vary.

Lesson 17

Problem Set

Line drawn from each butterfly to each flower

Plates with matching number of apples drawn

Homework

Line drawn from each pail to each shovel

Line drawn from each plate to each fork; 1 fork drawn

4 fishbowls drawn

Lesson 18

Problem Set

Line drawn from each hat to a scarf; scarves circled; X placed on two scarves

More than 6 leaves drawn

Homework

Line drawn from each hoop to a ball; ball circled; 6

8; 8; *YES* circled

Lesson 19

Problem Set

Set of 5 ladybugs circled; 2 ladybugs drawn

Set of 6 watermelon slices circled; 1 watermelon slice drawn

Fewer suns than stars drawn

Homework

1 bird drawn

5 dogs drawn; fewer than 5 doghouses drawn; 5 bones drawn

Lesson 20

Sprint

1	2
2	2
3	1
3	3

Problem Set

2 beads colored; 5 beads colored; bottom chain circled; 5, 2

6 beads colored; 4 beads colored; top chain circled; 6, 4

Answers will vary; chain with fewer beads circled

Answers will vary; chain with fewer beads circled

More chains drawn; number recorded for each

Homework

3 beads colored blue; more than 3 beads colored red; answers will vary.

5 beads colored green; fewer than 5 beads colored yellow; answers will vary.

2 beads colored brown; more than 2 beads colored blue; answers will vary.

9 beads colored red; fewer than 9 beads colored green; answers will vary.

Chain with more than 3 beads but fewer than 10 beads drawn

Chain with fewer than 10 beads but more than 4 beads drawn

Lesson 21

Sprint

See Kindergarten Module 3 Lesson 20 answer key.

Recording Sheet

Answers may vary.

Problem Set

Circles colored red; 2

Triangles colored yellow; 5

Hexagons colored green; 3

Rectangles colored orange; 9

Rectangle colored

Triangle colored

Rectangle colored

Triangle colored

Circle colored

Circle colored

Homework

Set of bicycles circled

Set of children circled

Set of moons circled

5 books drawn; some apples drawn; answers may vary.

Lesson 22

Problem Set

5 circles drawn;

7 circles drawn;

10 circles drawn;

Answers will vary.

Homework

8 nests drawn;

6 trees drawn;

10 bananas drawn;

Some pencils and an equal number of crayons drawn

Lesson 23

Problem Set

3; 4 leaves drawn, 4

5; 6 fish drawn, 6

7; 8 acorns drawn, 8

9; 10 pieces of corn drawn, 10

Answers will vary.

Homework

6; 7 balls drawn; 7

9; 10 peanuts drawn; 10

Lesson 24

Problem Set

5; 4 suns drawn, 4

7; 6 clouds drawn, 6

9; 8 sharks drawn, 8

10; 9 worms drawn, 9

Answers will vary.

Homework

5; 4 circles drawn, 4

7; 6 circles drawn, 6

10; 9 circles drawn, 9

9; 8 circles drawn, 8

Lesson 25

Sprint

See Lesson 20 answer key.

Problem Set

3; 5; 5; 3

7; 8; 8; 7

8; 10; 10; 8

Answers will vary.

Homework

5; 6; 6; 5

7; 5; 7; 5

5; 6; 6; 5

Lesson 26

Problem Set

6; 4; 4; 6

5; 7; 5; 7

8; 9; 8; 9

Answers will vary.

Homework

9; 8; 8; 9

8; 10; 8; 10

7; 9; 7; 9

Lesson 27

Problem Set

Tower with more than 3 cubes drawn; answers will vary.

Tower with fewer than 5 cubes drawn; answers will vary.

Tower with more than 7 cubes drawn; answers will vary.

Train and another train with fewer cubes drawn; answers will vary.

Tower and another tower with fewer cubes drawn

Homework

Tower with more than 5 cubes drawn; answers will vary.

Tower with fewer than 9 cubes drawn; answers will vary.

Train with more than 7 cubes drawn; answers will vary.

Tower and another tower with more cubes drawn; answers will vary.

Lesson 28

Sprint

3	3
4	5
5	1
4	3

Problem Set

5; 3; 3; 5

7; 6; 6; 7

8; 6; 6; 8

10; 9; 9; 10

Answers will vary.

Homework

7; 4; 4; 7

10; 9; 9; 10

8; 6; 6; 8

5; 4; 4; 5

Answers will vary.

Lesson 29

Recording Sheet

Answers will vary.

Homework

Line drawn from 4 full containers to *Full*

Line drawn from 3 partially full containers to *Not Full*

Line drawn from 3 empty containers to *Empty*

Lesson 30

Recording Sheet

Answers will vary.

Homework

Apples colored per directions

4; 6

2; 8

7; 3

1; 9

9; 1

3; 7

Lesson 31

Sprint

3	5
4	4
5	3
3	5
4	5
5	4

Problem Set

Imaginary animal composed of various shapes drawn

Homework

Castle composed of various shapes drawn

Lesson 32

Recording Sheet

Answers will vary.

Homework

Answers will vary.